T0296619

The Neuroscience of Meditation

Understanding Individual Differences

The Neuroscience of Meditation

Understanding Individual Differences

Yi-Yuan Tang
Department of Psychological Sciences, Texas Tech University,
Lubbock, TX, United States

Rongxiang Tang
Department of Psychological & Brain Sciences,
Washington University in St. Louis, St. Louis,
MO, United States

ACADEMIC PRESS

An imprint of Elsevier

Academic Press is an imprint of Elsevier
125 London Wall, London EC2Y 5AS, United Kingdom
525 B Street, Suite 1650, San Diego, CA 92101, United States
50 Hampshire Street, 5th Floor, Cambridge, MA 02139, United States
The Boulevard, Langford Lane, Kidlington, Oxford OX5 1GB, United Kingdom

Notices
Knowledge and best practice in this field are constantly changing. As new research and experience broaden our
understanding, changes in research methods, professional practices, or medical treatment may become
necessary.

Practitioners and researchers must always rely on their own experience and knowledge in evaluating and using
any information, methods, compounds, or experiments described herein. In using such information or methods
they should be mindful of their own safety and the safety of others, including parties for whom they have a
professional responsibility.

To the fullest extent of the law, neither the Publisher nor the authors, contributors, or editors, assume any
liability for any injury and/or damage to persons or property as a matter of products liability, negligence or
otherwise, or from any use or operation of any methods, products, instructions, or ideas contained in the
material herein.

British Library Cataloguing-in-Publication Data
A catalogue record for this book is available from the British Library

Library of Congress Cataloging-in-Publication Data
A catalog record for this book is available from the Library of Congress

ISBN: 978-0-12-818266-6

For Information on all Academic Press publications
visit our website at https://www.elsevier.com/books-and-journals

Publisher: Nikki Levy
Acquisitions Editor: Melanie Tucker
Editorial Project Manager: Pat Gonzalez
Production Project Manager: Bharatwaj Varatharajan
Cover Designer: Matthew Limbert

Typeset by MPS Limited, Chennai, India

Working together
to grow libraries in
developing countries

www.elsevier.com • www.bookaid.org

Contents

Preface

Meditation is an ancient contemplative practice and has primarily served the purpose of alleviating mental suffering in traditions such as Hinduism and Buddhism throughout history. However, meditation can be difficult to define because practices vary between traditions and within them. In general, meditation is a mind and body practice that aims to train attention and awareness, increase relaxation and calmness, improve psychological balance, and enhance overall health and well-being (Goleman, 1988; NCCIH, 2016; Tang, 2017; Walsh & Shapiro, 2006). By introspectively attending to, and becoming more aware of, one's thoughts, emotions, and sensations that arise at any moment, meditation seeks to foster a nonjudgmental, nonreactive, and open attitude toward everyday experiences, irrespective of their pleasantness or unpleasantness (Tang, 2017; Tang, Hölzel, & Posner, 2015). Based on nearly three decades of scientific research and clinical application, meditation practices have been shown to promote psychological health and well-being, as well as certain aspects of cognitive function and physiological health (Brown & Ryan, 2003; Goyal et al., 2014; Tang et al., 2015). Interestingly, meditation was not originally intended to be practiced among the general public for the purpose of improving health or function; instead, it was practiced within religious settings among yogis and monks for the purpose of alleviating suffering and attaining spiritual enlightenment (Goenka, 2003; Hart, 2009).

However, as time goes on, all ideas and traditions would experience some extent of development or different perceptions even if they are intended to stay unchanged. As human beings are similar to one another in many ways, the desire to be happy and free of suffering can be found outside the religious realms of Buddhism and Hinduism. Daily stress and emotional ups and downs are common and extremely costly for mind and body, prompting many individuals to seek cures that could help them handle the constant slings and arrows of life. Indeed, unlike hundreds of years ago when the growth of the global population and living pressures were not as tremendous as they are nowadays, people in contemporary society do seem to experience more sources of pressure and competition at many different levels. For instance, technological advances in the form of computers, cellphones, and social media allow us to connect with the world faster and more efficiently, but they also generate sources of pressure and burden on our body and mind, one example being required to process and respond to the extra information in a timely manner. Naturally, given these various daily stressors, meditation attracted public attention and became popular among people who do not have a religious interest. Since the 1970s, meditation has not only gained popularity, but also undergone considerable development and adaptation to better suit the needs of contemporary society (Kabat-Zinn, 2013; Maharishi, 2001). Consequently, different meditation techniques and practices were adapted and incorporated into secular and standardized programs (e.g., transcendental meditation, mindfulness-based stress reduction) for various purposes such as stress reduction and treating mood disorder symptoms. These programs are taught to anyone who would like to learn and utilize meditation in their daily life. It is fair to say that this movement of secularizing meditation practices has been very successful, and meditation is increasingly becoming a part of a healthy lifestyle, almost equivalent to the status of physical exercise in people's daily routine. Currently, nonsecular meditation practices and techniques are being actively practiced by people around the world (National Health Interview Survey (NHIS) Report, 2017).

The burgeoning popularity of meditation has triggered extensive scientific research into this promising mental training technique, which includes investigation into different meditation techniques from secular and religious practices. While many empirical studies have found a wide array of benefits associated with meditation practices, long-term and short-term, most of the studies generalized the effects and benefits of meditation to the population level, meaning that they tended to draw the conclusion that anyone who goes through the same kind and length of practice would obtain equivalent effects on health and well-being. This kind of generalization is not totally invalid within a scientific context because scientists often make inferences about a population based on observations of various subgroups or samples of a population. However, generalizations can be problematic if we aim to understand individual differences. One may already start to wonder why someone would want to investigate individual differences in the first place, especially in the context of meditation practices. There are many reasons why it is important to focus on the role of individual differences in influencing the actual learning, practice, and outcomes of meditation. A discussion of these reasons is a key theme throughout this book, but as one example, considering individual differences would increase the effectiveness and cost-effectiveness of meditation in clinical application settings. Meditation has been widely applied within clinical contexts for treating the psychological symptoms of certain disorders, indicating that achieving the maximum effectiveness for *each individual* would be highly critical. Second, advancing our existing knowledge of meditation effects and mechanisms through investigation of individual differences would also be extremely valuable for identifying people for whom meditation may not be suitable.

In fact, as personality psychology has already informed us about the impacts of individual differences in influencing our behavior

(Leary & Hoyle, 2009), it is not unreasonable to expect that the practice and effects of meditation may also be affected by such differences. For example, research of other similar forms of mental training and psychological intervention has suggested evidence of individual variability with regard to intervention effectiveness. Additionally, given the diverse nature of meditation techniques and practices themselves, it is highly plausible that different techniques and practices may have distinct effects at the individual level and that individuals may find one technique or practice more helpful than alternatives. From a research perspective, there is still a lot to be learned about meditation. As the field has witnessed both exciting and conflicting research findings with regard to the positive effects of meditation, considering individual differences may be one of the ways to answer the question of why meditation seems to exhibit mixed effects across studies. This is not to suggest that mixed findings are solely the result of overlooking individual differences, instead it is critical to evaluate all possible sources of variances that may contribute to different experimental observations. Moreover, the lack of knowledge of individual differences in meditation would undoubtedly move us away from comprehensively understanding its mechanisms and effects. From a clinical standpoint, attending to individual differences when providing a psychological intervention technique such as a meditation program to a patient would allow for more precise and even more personalized approaches to maximize treatment efficacy. Recently there has been increasing advocacy for precision medicine and personalized medicine from the US government, highlighting a societal level of awareness and emphasis on the importance of considering individual differences in clinical settings. Therefore understanding the role of individual differences in meditation is highly useful and informative for future scientific investigation of this growing research field, as well as for healthcare practitioners and the general public who are deeply invested in gaining health-related benefits from practicing meditation.

Driven by these motivations and considerations of advancing our scientific understanding and clinical practices of meditation, this book is meant to provide a better understanding of meditation and its benefits from a unique, individual differences perspective based on existing scientific theories and research. Special emphasis is given to the rapidly growing neuroscientific findings of meditation. Previously, individual differences in meditation have not received much attention, thus empirical research focusing on this topic has largely been preliminary or exploratory in nature. In this book, we review and discuss major theoretical and conceptual frameworks of individual differences in relation to meditation practices and describe existing behavioral and neuroimaging studies that either have looked at, or are related to, the role of individual differences in response to meditation. It should be noted that most of the evidence supporting our discussion consists of relevant individual differences research on similar psychological interventions. We identify various sources of individual differences relevant for meditation practices based on research from biology, genetics, culture, development, personality, and neuroscience. In order to encourage further discussion and investigation across different disciplines, we offer suggestions for potential research topics and useful methodologies. We pay close attention to several important conceptual and methodological issues concerning meditation research raised by researchers from the field. Finally, we end the book with an outlook for the future of meditation research, its translational application, and its utility in a public setting.

Outline of the book

The book has 10 chapters and is organized to include experimental studies and established scientific theories from education, biology, psychology, and neuroscience. We include a special section on the implications for translational research and application in each chapter in order to demonstrate the practical utility of the discussed theme. The word *meditation* as it appears in this book broadly includes both secular and religious meditation practices. Hence different meditation programs, techniques, and traditions are clearly named throughout the book when applicable. A brief introduction and summary of the content of each chapter follows next.

Chapter 1: Theoretical frameworks of individual differences in interventions

In the chapter we seek to build a theoretical foundation for examining and considering individual differences in meditation research and application by discussing prominent conceptual frameworks from psychology and psychotherapy. We primarily focus on two notable frameworks, namely patient-centered care and the aptitude (or attribute) by treatment interaction (ATI) paradigm. Patient-centered care is a prevalent conceptual model within medical settings and stresses the importance of acknowledging and honoring individual preferences and needs in treatment. It has been widely embraced by healthcare providers and governments as a guiding principle for providing high quality healthcare services. The ATI framework emphasizes the notion that treatment outcomes depend on the interaction between individual attributes and the chosen intervention. Based on this framework, individual differences such as personality traits are considered to be attributes that could influence the effects of intervention and have been extensively investigated and taken into consideration in psychotherapy; especially when matching patients to the appropriate treatment approaches. We unfold our discussion around these two conceptual frameworks, but also touch upon other relevant governmental initiatives such as precision medicine and personalized medicine that underscore the need for considering individual variability in different psychological interventions. We highlight specific connections and implications these frameworks have for meditation research and application.

Chapter 2: Personality and meditation

The chapter includes a discussion of how personality traits and other individual characteristics could influence meditation practices in terms of practice engagement, frequency, and the effectiveness of achieving the desirable outcomes. There is a wide range of individual traits and characteristics that have been studied to date, which makes covering all of them implausible. We will focus on the big five personality framework as a starting point to discuss how the five personality dimensions are related to meditation based on existing scientific evidence. For example, studies have demonstrated that people with higher levels of openness to experience tend to engage more in contemplative practices such as meditation and are more likely to practice outside of a training context in their daily lives. Similarly as for other forms of psychotherapy, individual differences in conscientiousness and extraversion also have specific impacts on how people react to interventions and adhere to intervention instructions, thereby affecting the extent of intervention benefits after completion. In an attempt to be more comprehensive when reviewing relevant individual characteristics and traits, we describe individual differences in personal goals, motivations, and values pertinent to intervention effectiveness and outcomes. Our goals are to highlight possible investigative targets for future individual differences research in meditation and offer a comprehensive overview of how individual differences in personality traits and characteristics contribute to differential efficacy following psychological interventions, especially meditation, across individuals.

Chapter 3: Cultural differences in meditation

It is widely known and established through daily observation and scientific research that people from Western and Eastern civilizations hold different thoughts, values, and beliefs, which could affect meditation practice and outcomes. There has been ample evidence indicating that people from different cultures—Chinese and American cultures are exemplars—perceive and approach the world differently, pay attention to, and make decisions differently, given the same scenarios. For example, people from American and Chinese cultures show different emotional processes, social behavior, and self-related processes (e.g., collectivism vs individualism or independence vs interdependence) and utilize distinct neural pathways for mental processing such as attention or simple cognitive tasks. Cultural differences can affect contemplative practices such as meditation given that meditation is rooted in cultures and traditions, and is a complex process of attention, emotion, cognition, decision making, and social interaction. In this chapter we describe cultural differences in diverse domains, especially with regard to how these macro-level differences may impact the actual learning and practice of meditation, its potential benefits in well-being and health, as well as symptoms and disorders. We focus on discussing these differences in both behavioral and neural perspectives between Eastern and Western cultures in meditation and how such differences may affect the outcomes of meditation. To conclude the chapter we discuss a cross-cultural health model and explore how to develop culturally adapted preventions and interventions (such as meditation) to promote behavior change, health, and well-being.

Chapter 4: Genetic association with meditation learning and practice (outcomes)

The chapter adds another layer of complexity to the individual differences discussion of meditation by considering biological factors that may govern the learning and practice experiences of meditation. Given the growing emphasis on genes by environment interaction from notable biomedical research models, we focus on this rarely discussed topic in the research of psychological intervention—how genetic predisposition could affect how well individuals learn and respond to different ways of teaching, particularly within intervention programs. Currently not much is known about the precise mechanisms through which genes interact with mind and body to affect behaviors such as learning. Thus we propose a putative framework to describe how we think genes may come into play when individuals first learn to practice meditation. Furthermore we discuss how genes may be shaped by meditation experiences by highlighting recent work on how meditation experiences, short-term and long-term, induce epigenetic changes in adults. These findings could be promising for understanding how meditation practices could have a lasting imprint on genes, while illustrating the interplay between nature and nurture. To conclude the chapter, potential genetic associations with personality traits are briefly discussed.

Chapter 5: Sympathetic and parasympathetic systems in meditation

Meditation practices engage people's sympathetic and parasympathetic systems to induce a calm and effortless state of consciousness. It also requires a balanced coordination of these two systems to maintain a meditative state throughout the duration of practice. In this chapter we extensively discuss a theoretical and mechanistic framework of how these two systems are engaged in meditation based on existing empirical evidence. We further discuss how these two systems are subsequently altered by short-term and long-term practice experiences. We suggest that different depths of meditative states and stages of practice expertise may manifest in both systems and have distinct biological or physiological signatures. Prior work has shown various levels of changes during and following meditation practices in the parasympathetic system such as breathing amplitude, breathing rate, and skin conductance response, as well as in the sympathetic system, predominately involving the brain. We review these lines of work throughout our discussion with an emphasis on the interaction between the two systems. We also describe potential individual differences in these two systems that may play a role in influencing the practice effects and outcomes of meditation. Although we touch upon some of the empirical work involving the brain, this chapter serves as a primer for Chapter 6, Brain regions and networks in meditation, which entails a detailed discussion of how the brain is involved in meditation.

Chapter 6: Brain regions and networks in meditation

Neuroscientific studies on meditation are gaining momentum. We first describe critical brain regions commonly involved in meditation and their respective functions and roles in inducing and affecting meditative states and outcomes. However, as the brain is a complex system operating via multiple networks and regions, we expand our discussion to identify key brain networks associated with meditation practices and effects on psychological health and cognitive function. We focus on three networks: default mode network, cinguluo-opercular network (sometimes called the salience network), and the frontoparietal network (executive control network). Each of these networks includes brain regions that have previously been identified separately in neuroimaging studies on meditation, but a system-level connection has not often been discussed or investigated. We highlight recent advances in cognitive neuroscience methodologies and techniques that capture the network level properties of the brain and explain how these strategies could be useful for meditation research, especially for examining individual differences in brain changes following meditation. We also describe some of the promising studies that have started to examine network level changes following meditation. We conclude by suggesting that looking into the brain for putative mechanisms of meditation on how it improves the functions of mind and body could be a very promising endeavor.

Chapter 7: Meditation over the lifespan

In the chapter we describe the practice of meditation among different age groups and present relevant scientific findings on the wide range of practice benefits detected across all ages. To demonstrate meditation as a contemplative practice that can be engaged over the course of one's lifespan, we begin our discussion with the implementation of meditation programs in education settings for children and adolescents. For instance, meditation programs have been widely implemented in the United Kingdom for school children to practice on a regular basis. Additionally, there has been some promising evidence suggesting that meditation could be useful for the well-being and academic performance of young adults in schools. We then turn our focus on meditation practices in middle-aged adults and those from the older population who have also actively participated in such mental training techniques. As meditation is increasingly becoming a lifestyle practice and intervention, its role and effects across development and over lifespans are worth investigating longitudinally, not just limited to a specific age group only for a period of time. We evaluate the potential of meditation as mental training for promoting psychological health and well-being, and whether or not it would be suitable for all age groups. Last, we suggest that considering individual differences is important when making such evaluations, as individuals with specific personality traits and characteristics may find meditation more appealing and beneficial, which may motivate them to maintain the practice over their entire lifespan.

Chapter 8: How to measure outcomes and individual differences in meditation

Investigating individual differences requires careful consideration of experimental designs and tools of assessment. This chapter provides a primer on how to best examine and assess individual differences in meditation effects and outcomes. We start by describing how meditation effects and outcomes have generally been measured in existing literature, which have mostly been group studies that did not consider individual differences. We then move onto specific discussion of assessing individual differences in frequently targeted outcomes that tap into psychological well-being, cognitive function, and physiological health. We include a few empirical examples that have adequately characterized individual differences in meditation as a part of our discussion, while also considering methodological constraints unique to meditation research that may impede individual differences investigations. The scientific community has paid increasing attention to research designs and methodologies, as well as common methodological issues that can become problematic for ensuring the validity and reliability of research findings. We not only suggest promising and rigorous paradigms and tools sensitive to individual differences for future investigation of meditation, but also discuss major methodological challenges in meditation research, highlighting the need for caution when designing experiments and interpreting results. Last, to make our discussion more concrete, we offer some interesting research ideas and examples of studying individual differences within the context of meditation.

Chapter 9: Personalized meditation

In this chapter we present the concept of personalized meditation to advocate an individualized approach in the implementation of meditation programs in public and in clinical settings. The proposed idea is different from the concept of precision medicine, in that it specially seeks to develop individualized programs, rather than choosing from existing programs. Meditation programs have largely been standardized so that every individual receives the same curriculum, often taught by the same instructor. However, individual variability in response to meditation is evident and worth taking into consideration in teaching and in practices. We discuss several ways through which meditation can be personalized to better suit the needs and preferences of each individual, while concurrently maximizing meditation-related benefits in enhancing health and well-being. One of the major factors to be discussed when planning personalized meditation for individuals are personality traits and characteristics. We also describe how personal preferences for different teaching styles and modalities could also be utilized to further improve the overall personalized meditation experiences. We conclude by suggesting that attending to individual needs and preferences through a personalized meditation program or curriculum is a potentially promising approach for increasing the overall effectiveness and cost-effectiveness of meditation.

Chapter 10: Critical questions and future directions in meditation

In this chapter we summarize major topics and issues discussed in previous chapters and suggest promising future directions for examining meditation effects and benefits, as well as unsolved research questions concerning individual differences in meditation. In particular, the following topics are discussed: (1) the importance of instructor on influencing meditation outcomes; (2) the influence of different meditation techniques, stages of expertise, and amount of effort used on meditation outcomes; (3) the feasibility of combining meditation with other training approaches to enhance targeted outcomes; and (4) the development and potential of meditation in the digital age. We would like to encourage discussion across different research disciplines and people from all interested communities on utilizing technological advances to promote the effects of meditation. We would like to brainstorm different ways of offering meditation practices through technology such as mobile apps and online platforms. Another interesting prospect that is discussed with regard to technology is the use of real-time feedback to monitor and improve the quality of meditation practices at the individual level, which has been increasingly tested in recent years. We also discuss the potential side effects or adverse effects if someone does not practice meditation properly. In conclusion we suggest that research on meditation still holds immense promise and potential for revealing fascinating facts and knowledge about the precise mechanisms of contemplative practices underlying improvement in health and well-being.

References

Brown, K. W., & Ryan, R. M. (2003). The benefits of being present: Mindfulness and its role in psychological well-being. *Journal of Personality and Social Psychology*, *84*, 822–848.

Goenka, S. N. (2003). *Meditation now: Inner peace through inner wisdom*. Pariyatti Publishing.

Goleman, D. (1988). *The meditative mind: The varieties of meditative experience*. New York: Tarcher.

Goyal, M., Singh, S., Sibinga, E. M., Gould, N. F., Rowland-Seymour, A., Sharma, R., ... Haythornthwaite, J. A. (2014). Meditation programs for psychological stress and well-being: A systematic review and meta-analysis. *JAMA Internal Medicine*, *174*(3), 357–368.

Hart, W. (2009). *Art of living: Vipassana meditation*. HarperOne.

Kabat-Zinn, J. (2013). *Full catastrophe living (revised edition): Using the wisdom of your body and mind to face stress, pain, and illness*. Bantam.

Leary, M. R., & Hoyle, R. H. (2009). *Handbook of individual differences in social behavior*. Guilford Press.

Maharishi, M. (2001). *Science of being and art of living: Transcendental editation*. Plume.

National Health Interview Survey (NHIS) Report. (2017). *More adults and children are using yoga and meditation: Nationwide survey reveals significant increases*. NCCIH Press Office.

Tang, Y. Y. (2017). *The neuroscience of mindfulness meditation: How the body and mind work together to change our behavior*. Cham: Springer Nature, Meditation: In Depth. NCCIH. 2016.

Tang, Y. Y., Hölzel, B. K., & Posner, M. I. (2015). The neuroscience of mindfulness meditation. *Nature Reviews Neuroscience*, *16*, 213–225.

Walsh, R., & Shapiro, S. L. (2006). The meeting of meditative disciplines and western psychology: A mutually enriching dialogue. *American Psychologist*, *61*(3), 227–239.

Chapter 1

Theoretical frameworks of individual differences in interventions

Individual differences are ubiquitous in everyday life, but are often overlooked when medication, intervention, or treatment is deemed suitable and appropriate for the general population. As such, individual needs and preferences are often overlooked by the individuals themselves and those providing the treatment. Indeed, when we operate under the assumption that an intervention works for everyone, we unintentionally fall into the trap of thinking a universal panacea exists for the targeted symptoms and would work the same for different individuals. In reality, people are far more complex and diverse, which can lead to differential responsiveness to the same treatment approach, subsequently affecting the extent of the treatment's effects on the targeted outcomes. For instance, even for medication (e.g., penicillin) that commonly work for most of the population, exceptions such as drug allergy and nonresponsiveness occur in a small percentage of people. When it comes to psychological interventions we are even more likely to not think about whether our unique preferences and characteristics would make us less suitable for, and less responsive to, a specific intervention. Aside from clinical applications, the presence of individual differences can also be found in scientific investigations. However, individual variability is often not carefully considered when it comes to empirical studies of psychological interventions, as most studies only examine group-level effects of the intervention on specific outcomes and assume it is equally effective and suitable for everyone within the studied group.

As one kind of group-based psychological interventions for improving psychological symptoms and health, meditation has been actively applied and practiced within clinical settings. It is also being widely practiced as a mental training or lifestyle intervention by people from the healthy population who seek to enhance psychological well-being and cognitive function. Similar to other kinds of psychological interventions, the benefits of meditation have been mostly generalized to the population level in prior research, yet mixed outcomes are increasingly being reported by studies

The Neuroscience of Meditation. DOI: https://doi.org/10.1016/B978-0-12-818266-6.00002-2

(MacCoon et al., 2012; MacCoon, MacLean, Davidson, Saron, & Lutz, 2014; Rosenkranz et al., 2013). For instance, the positive effects on sustained attention were not significant when comparing the mindfulness-based stress reduction (MBSR) (a meditation-based intervention) with an active control intervention (MacCoon et al., 2012). Likewise, the MBSR program does not seem to be more effective than active control intervention in reducing psychological distress and physiological symptoms (Rosenkranz et al., 2013). Typically, when examining the effects of an intervention, a randomized controlled trial design with an active control intervention is used for the experiment so that we are able to compare our intervention effects against that of a similar alternative in order to minimize potential confounding variables. The benefits in enhancing psychological well-being and attention were not replicated as in other studies that also utilized active control interventions for comparison. Looking at the effect sizes of meditation studies, several meta-analyses of different studies on meditation benefits have shown that the degree of improvement in psychological health and aspects of cognitive function following meditation have varied from small to moderate (Chiesa, Calati, & Serretti, 2011; Gu, Strauss, Bond, & Cavanagh, 2015), suggesting that there is a considerable amount of variability across studies.

These mixed findings point to two possible explanations: (1) meditation does not have an effect on individual well-being and cognition (or only has a subtle effect); and (2) meditation does not induce the same level of benefits in every individual and the extent of improvement may be affected by preexisting individual differences. The first explanation has been proposed by skeptics, who think that the observed positive effects of meditation could either be spurious findings or an exaggeration of subtle benefits resulting from meditation practices. Regardless of which explanation may be true, the implications for clinical practices and recreational practices of meditation are profound as we would not want to waste resources and time on an intervention that is not going to be effective or anywhere better than the existing programs for promoting health and well-being. If we chose the first explanation, we have to rule out the possibility that mixed findings may be due to neglecting individual differences in the observed effects. Because group-averaged results are largely affected by the study sample—and it is possible that some study samples had individuals who were more responsive to the intervention than other study samples—this can lead to subtle or no effects on the outcomes of interest. Unfortunately, there has been a lack of knowledge regarding individual differences in response to meditation or on individual factors associated with intervention effectiveness, making it difficult to confirm either of the possibilities. Even when individual differences have been considered in the past, the study results were mostly preliminary and have yet to reach a wider audience for the studies to have a substantial influence on the scientific community and the public. Hence increasing awareness of these issues and considering individual differences are crucial for scientific investigation to accurately evaluate the effects of meditation and for clinical application of meditation practices in treating problematic symptoms to be more effective.

We will introduce the theoretical background of individual differences in treatment responsiveness and its relation to meditation research and application. To facilitate better understanding of the topic, we start by describing two notable conceptual models and frameworks particularly relevant for considering individual differences in terms of treatment planning and treatment effectiveness, namely (1) patient-centered care, and (2) aptitude (or attribute) × treatment interaction. We also discuss precision medicine, a national initiative proposed by the US government, which further demonstrates the growing societal awareness for individual variability in response to treatments, as well as governmental support for scientific research in this exciting area.

Patient-centered care

Within a medical setting healthcare services are typically provided at the level of individual, where physicians and other professionals closely interact with the patient to discuss a treatment plan and offer the necessary services. It is therefore not surprising that individual differences and preferences are widely respected and valued within healthcare systems. The patient-centered care approach is a prominent example advocating the importance of attending to individuals' preferences, values, characteristics, and needs when providing treatment and care to patients (IOM, 2001; Rathert, Wyrwich, & Boren, 2013). It underscores the idea that patients should be at the center of the treatment process, not the middle, and should also be able to make decisions regarding their treatment and have their preferences respected and honored whenever possible (Steiger & Balog, 2010). Notably, patient-centered care does not limit the forms of treatment only to medical procedures, it is also intended to emphasize the value of respecting individuality and differences and being responsive to individual needs across any form of treatment and treatment context. This approach is relevant for psychological interventions such as cognitive behavioral therapy (CBT) and meditation-based programs that are often used within clinical settings for treating symptoms of psychiatric disorders. However, it is important to realize that the concept of patient-centered care has been around for more than 50 years and has been widely embraced by physicians, clinicians, healthcare organizations, and governments. Since 2001, the Institute of Medicine has included patient-centered care as one of the six critical aims for improving the health care system in the 21st century (IOM, 2001).

As one of the key objectives for improving the quality of health care, the philosophy of patient-centered care has been gradually realized in medical practices and has been quite successful in terms of patients' satisfaction, relation to treatment providers, and treatment outcomes (Epstein & Street, 2011; Rathert et al., 2013; Rocco, Scher, Basberg, Yalamanchi, & Baker-Genaw, 2011). For instance, a systematic review of patient-centered care indicated that nearly all reviewed studies showed a positive relationship between patient-centered care and the patient's satisfaction and well-being (Rathert et al., 2013). Although the impacts of patient-centered care on long-term clinical outcomes still require more thorough investigation, an overall positive trend in different outcome measures can be widely observed in most studies. The idea, however, was not without pushbacks. The debate between proponents and skeptics on whether focusing on individual needs and preferences may contradict the purpose of evidence-based treatment that is typically standardized for the general population eventually came to the conclusion that when it comes to treatment, one should consider "both the art of generalizations and the science of particulars" (Epstein & Street, 2011). The philosophy of patient-centered care has profound impacts on how we think about standardized treatment programs when providing them to individuals and how we should evaluate the outcomes of these treatment programs at the level of individuals. However, it is important to recognize that the concept of patient-centered care is easier to understand than to implement in practice, since providing patient-centered care often requires infrastructural changes to the existing healthcare facilities, as well as adequate personnel training in order to offer high-quality treatment and services (Epstein & Street, 2011). Otherwise the devoted efforts to pursue patient-centered care may appear superficial and merely theoretical. For the purpose of our discussion the conceptual implications of patient-centered care for psychological interventions is much more important than the implementation aspect of this framework.

Psychological interventions are also considered within the patient-centered care framework. When one thinks about psychological intervention or psychotherapy that does not involve medication, the image that often comes into mind is an individual-based interaction with the treatment provider, typically a clinical psychologist or therapist who devotes undivided attention to the patient and designs treatment plans specifically tailored to the needs of the individual. Within the field of psychotherapy there has always been an emphasis on individual needs and characteristics, which is consistent with the philosophy of patient-centered care (Epstein & Street, 2011). Although not all psychological interventions or therapies are focused on the individual level, group-based interventions such as meditation and CBT could also benefit from this concept of patient-centeredness. In particular, thinking through the perspective of individual patients, instructors can at least become more aware of, and attentive to, potential individual differences in needs and preferences within a group, rather than assuming that everyone in the group would share the same values and goals for the intervention. Conceptually, cherishing the value of patient-centeredness when providing psychological interventions is important for ensuring the effectiveness, satisfaction, and quality of the intervention—as is the case for medical settings.

Speaking of group-based interventions such as meditation, there has been a movement of integrating patient-centered care philosophy within complementary and alternative medicine to increase the quality and outcome of overall treatment (Maizes, Rakel, & Niemiec, 2009). Complementary and alternative medicine approaches such meditation-based interventions and other nonconventional medical therapies are frequently provided in conjunction with conventional medical treatments. Therefore, valuing and respecting individual differences in the application of meditation-based interventions would undoubtedly be pivotal for ensuring high-quality intervention. The evaluation of intervention effects could also be informed by the philosophy of patient-centered care. For instance, one encouraging work on meditation-based intervention chose to examine specific patient-centered outcomes such as self-efficacy, work productivity, and health services utilization after 1 year of intervention, rather than simply looking at conventional measures of psychological well-being (McCubbin et al., 2014). The study found that meditation-based intervention increased self-efficacy and decreased the use of health services, suggesting that such intervention is effective and that there are alternative possibilities for evaluating intervention outcomes, especially at the individual level. It also suggests examining individual-specific measures could be appropriate for future studies and potentially more sensitive for capturing improvement in well-being following interventions. Overall, the concept of patient-centered care is transforming our healthcare system and has far-reaching implications for informing both scientific research and clinical practices.

Aptitude (or attribute) × treatment interaction

We primarily discussed the philosophy of patient-centered care from a conceptual standpoint and described its promising potential for improving the outcomes of intervention. What is missing from this framework is how exactly it could be translated into helping scientific investigations to accurately evaluate intervention outcomes, since patient-centered care focuses more on improving the quality of clinical practices and intervention outcomes. However, if we look into the field of psychotherapy there is a notable conceptual framework geared toward studying individual differences in response to psychological intervention.

The aptitude (or attribute) x treatment interaction (ATI) was a research paradigm and conceptual framework proposed by Lee Cronbach in the early 1950s that linked individual differences to treatment outcomes. In this paradigm, aptitude (or attribute) refers to any relevant individual difference variables identified at the baseline (prior to any treatment or intervention), such as personality traits, motivation, attitudes, and beliefs that may interact with a chosen treatment (or intervention) to produce differential treatment effects on outcomes (Caspi & Bell, 2004; Cronbach, 1957). It should be noted that the term "interaction" within the ATI framework is used in a statistical sense, implying a moderating effect of individual attributes on the relationship between treatment and outcomes. From a practical standpoint the paradigm seeks to examine whether the chosen treatment or intervention is a match (or mismatch) for the patient based on his or her specific characteristics (Caspi & Bell, 2004). This innovative framework was one of the first calls for scientific investigations (not just clinical applications) to systematically take into account individual differences in the evaluation of treatment effects and consider the complex interactions between "organismic and treatment variables" (Cronbach, 1957; Snow, 1991). More importantly, the framework discourages researchers from relying on only examining the main effects of interventions by simply averaging any changes in outcomes across individuals within a group. That is, rather than generalizing the effects of treatment on outcomes to everyone, the ATI framework underscores the need to also examine how individuals respond to a chosen intervention given his or her specific characteristics, which subsequently affects the outcomes of the treatment. Hence the research question is not only "which treatment works the best," but also "which treatment works the best for whom" (Caspi & Bell, 2004).

When describing the ATI paradigm in relation to complementary and alternative medicine, Caspi and Bell (2004) suggested that the ATI paradigm can be used as a conceptual framework for conducting outcome research of different complementary and alternative medicine approaches. Yet, there seems to be a lack of awareness regarding the utility of the framework for the evaluation of intervention outcomes as prior research has mostly relied on group-averaged results to draw conclusions with regard to intervention effectiveness. We will delineate the major components of ATI framework and put it into the specific context of meditation research.

Starting with aptitudes, we briefly described these as any individual difference variables that could potentially moderate the relationship between intervention and outcome. A common misconception is that aptitudes only stand for personality traits or other fixed individual characteristics. However, aptitudes can refer to both fixed and dynamic constructs of individual variability across psychological, biological, and cultural dimensions (Caspi & Bell, 2004). Additionally, these attributes can have a reciprocal relationship with the interventions, such that they may also be affected over the course of intervention while exerting an influence over intervention outcomes. Nonetheless, choosing appropriate aptitudes for individual differences investigation can be tricky since there is a wide range of putative characteristics that may be relevant for our purposes. According to Caspi and Bell (2004) the main goal of ATI research is striving to be confirmatory rather than exploratory. Thus researchers should formulate a priori hypotheses regarding putative aptitude variables that are associated with differential intervention effects across individuals. Some of the early work inspired by the ATI framework mostly employed this hypothesis-driven approach due to concerns of potentially spurious statistical associations resulting from an exploratory "fishing" expedition. Because, simply by chance, one may detect a statistically significant yet spurious relationship between two variables if too many statistical tests were conducted with different pairs of variables. The term "treatment" in the ATI framework is more straightforward, referring to any intervention or psychotherapy that is provided to the individuals. In most cases, the treatment is a standardized intervention program or multifaceted package with various subcomponents that can be teased apart in order to examine them separately.

The term "interaction" refers to a statistical relationship between aptitudes and treatment that subsequently influences the extent of the treatment's effects on outcome measures. As Caspi and Bell described in their review of the ATI framework, there are two types of interaction, namely ordinal and disordinal. Let us consider a graph with the x-axis as aptitudes and the y-axis as outcomes. We have two regression lines representing two interventions with slopes that are not parallel. Ordinal interaction is when one intervention is always above the other intervention within the range of the aptitudes, even if they are not parallel, suggesting that one intervention has a superior effect on outcomes (Caspi & Bell, 2004). Disordinal interaction is when the two regression lines intersect at a specific point along of the x-axis, indicating that at a certain value of the aptitudes, the two interventions have the same effects on outcomes. However, except for this point of intersection, either intervention may be superior at different ranges of the aptitudes (Caspi & Bell, 2004). Therefore, it is crucial to always have at least two interventions in order to conduct ATI research.

By exploiting the strengths of the ATI paradigm, there have been some interesting empirical studies of differential intervention effectiveness across individuals (for a review, see Caspi & Bell, 2004). We will describe one example showing the utility and importance of the ATI paradigm in outcome research of psychological interventions. In a classic ATI study, Karno, Beutler, and Harwood (2002) sought to examine the effects of two interventions on improving alcoholism treatment outcomes. They selected emotional distress, externalizing coping style, and psychological reactance as their patient attributes (i.e., individual differences measures) and looked for the interaction between these attributes and two types of group-based intervention, CBT and family systems therapy (FST). The former can be considered as a close analog to meditation-based interventions. It is important to note that these individual attributes were chosen based on previous theoretical models hypothesizing their association with individual's response to psychotherapy; hence, the nature of the investigation is nonexploratory and theoretically oriented as advised by the ATI paradigm. Interestingly, the researchers found emotional distress to be the only attribute that moderated the effectiveness of CBT and FST in reducing alcohol use, such that CBT was more effective than FST for patients with low distress, but the effects of CBT diminished as distress levels increased (Karno et al., 2002). This study clearly demonstrated that individual variability influenced intervention effectiveness and that understanding such variability could inform the selection of a more appropriate intervention for each individual based on his or her characteristics.

This study also illustrated that when we seek to compare the effectiveness of two different interventions, simply comparing group-averaged effects on intervention outcomes may obscure systematic individual differences in response to a specific intervention (Shoham & Rohrbaugh, 1995). For instance, imagine a scenario where clinicians need to select the most appropriate and effective intervention for patients from two different options. They may be misguided by studies that overlook the role of individual variability and consequently choose an intervention that seems to have slightly higher effectiveness within specific groups. This unfortunate consequence would likely decrease the overall effectiveness and cost-effectiveness of the intervention programs, creating confusion regarding the actual effectiveness of the chosen intervention. Therefore, for studies on psychological interventions, including meditation, this ATI paradigm advocates for an individual differences perspective and provides useful guiding principles for evaluating the effectiveness of psychological interventions. However, unlike other similar kinds of psychological interventions (e.g., CBT), individual differences are not widely considered in studies of meditation. Hence building upon the existing framework and empirical research of ATI, future studies of meditation could focus on individual characteristics that have a theoretical bases for potentially predicting meditation effects on outcomes of interest to serve as aptitude variables in order for such investigations to be both feasible and meaningful.

Precision medicine initiative

Before we move onto identifying putative individual differences variables relevant for meditation effects on outcomes, it is worth taking the time to digress a little from the theoretical discussion to an introduction of an exciting ongoing health initiative proposed by the US government in 2015. The precision medicine initiative is a national effort for improving our healthcare systems and treatment approaches by emphasizing the importance of taking individual variability into account in clinical practice and scientific research (Collins & Varmus, 2015). This initiative began a new era for biomedical research and medicine in that it seeks to develop better and more effective treatment approaches for patients based on genetic, environmental, and lifestyle factors (National Research Council, 2011). In other words, the goal is to determine which treatment approaches would achieve the best outcome for which patients by taking individual characteristics into account. The major emphasis of precision medicine on individual variability is not new, as it clearly resonates with the two above mentioned theoretical frameworks in relation to individual differences in psychological interventions. However, the aim of precision medicine is much more ambitious. There are two primary components of the precision medicine initiative: (1) a short-term focus on cancer detection and treatment; and (2) a long-term focus on expanding knowledge applicable for health and a wider range of diseases (Collins & Varmus, 2015). The long-term goal is particularly relevant for our discussion on individual differences as it encourages scientists to consider different sources of individual variations in investigations and examine factors that play a role in influencing the degree of improvement in treatment or intervention outcomes. Behavioral factors such as personality traits have often been investigated in psychology and psychotherapy research, yet biological factors such as genetic variations are rarely considered, especially in terms of psychotherapeutic effects. The precision medicine initiative seeks to expand our horizon into different dimensions of human variability and comprehensively evaluate their contribution to human health and diseases. We will discuss how genetic predisposition can contribute to differential intervention effectiveness in Chapter 4, Genetic association with meditation learning and practice, but it is interesting to first take note of the lack of research on how genes influence our behavior and experiences within the intervention context.

Overall, the growing awareness of the importance of individual differences from governments, the public, scientists, and healthcare professionals propel promising advances in medicine and scientific research, which are unlikely to be successful without collaborative efforts across different disciplines and across nations.

Putative individual differences variables

There are many different sources of individual variability if one begins to think about how each person is unique. What is even more challenging is to determine which individual differences variables and factors are going to be the most relevant for our investigations. One should always begin the search of putative aptitude variables based on existing theoretical frameworks and empirical findings. In this section, rather than naming specific individual characteristics or traits, we aim to broadly discuss several key dimensions of individual differences from which investigators could pick their potential research targets based on their interests. We also tease apart each of these dimensions with more detailed discussion in subsequent chapters of this book, but would like to introduce them first in relation to the theoretical frameworks.

The first and most obvious dimension of individual variability is personality. We can easily observe different types of personality traits from people around us and can characterize them into distinct categories or label them using defined personality traits such as extraversion and conscientiousness. Thinking about individual differences in terms of personality traits are both intuitive and straightforward as there are several well-established personality frameworks for scientific investigations and readily available tools that can be used to measure these individual differences. For instance, the most influential personality framework is the five factor model or the "Big Five", namely extraversion, conscientiousness, openness to experience, agreeableness, and neuroticism. The Big Five personality traits can be measured by the well-validated NEO Five Factor Inventory, which has also been actively used across different fields of research as a standardized questionnaire for assessing personality. However, there are traits that not covered by the Big Five model but are relevant characteristics for intervention outcomes. For example, absorption is a personality trait referring to the capacity of immersing oneself into an experience (Caspi & Bell, 2004). Although not included in the Big Five model, absorption has been found to influence how well individuals respond to interventions for treating headaches (Neff, Blanchard, & Andrasik, 1983) and improving immune functionality (Gregerson, Roberts, & Amiri, 1996). These examples suggest that even though we would like to base our selection of aptitudes on well-established models of personality, there are other isolated traits with high conceptual relevance that should also be included in research.

The second source of individual differences are dynamic (or relatively malleable) characteristics that do not necessarily fall within the definition of personality traits. In our discussion of the ATI framework we stated that aptitude variables are not necessarily fixed and that dynamic constructs such as motivation and goals are also suitable for examining differential intervention effects across individuals. In particular, motivation and goals are imperative for intervention adherence and engagement, which have direct impacts on intervention outcomes (Mussell et al., 2000). This indicates that when we consider dynamic individual characteristics we should focus on factors related to learning and adherence as these two components are necessary for the intervention to exert an effect on the outcomes. Taking a meditation-based program as an example, if participants are not motivated to learn meditation practices or do not have the same goals as that of the program, then they are not likely to follow the instructions or even practice meditation outside of the program, consequently leading to no change or only subtle changes in the outcomes. An awareness of motivation and goals could help us select an intervention that best suits the needs of the participants. The bottom line is to have a logical hypothesis or convincing rationale for choosing certain characteristics as aptitude variables, rather than browsing through characteristics without a clear purpose.

The third source of individual differences is culture. Cultural differences are not readily apparent if one does not live in a multicultural society or travel and become exposed to different cultures. Indeed, back in the days when globalization was unimaginable, cross-culture interaction and communication were rare. However, we are now in a world where one can easily get from one continent to another continent within a matter of hours. This has led to cross-culture interaction, which also triggered the flourishing of meditation practices in Western civilizations since these practices were originally only known within Eastern cultures. Those who conduct experiments in North America, for example, are constantly dealing with a multicultural population and study samples often include individuals from various cultural backgrounds. Therefore by not considering how culture may shape individual differences, we would definitely be missing an important dimension of human variability. Cultural differences are coarser compared to personality traits and can manifest in different areas of our lives such as cognition, perception, and thinking. In order to tap into cultural differences, researchers could first compare intervention outcomes between two different cultural groups and then focus on specific outcome measures that are particularly affected by culture.

Summary and implications for translational research

Understanding individual differences in response to treatment and intervention is critical for health care and its future development. In this chapter we provided a detailed overview of existing theoretical frameworks and governmental initiatives that confirm the critical role of individual differences in influencing treatment effectiveness. We highlighted several dimensions of human variability to help investigators identifying appropriate variables for studies of individual differences in psychological interventions. Although these are by no means comprehensive and exhaustive, the goal is to reiterate the principle of choosing putative aptitude variables based on a priori hypotheses and existing research. We suggest that investigators should decide on one dimension of individual variability as a starting point when examining individual differences in intervention effectiveness for the first time. Conceptual themes from these established frameworks have already informed medical and clinical practices. Physicians and healthcare professionals are increasingly aware of individual differences presented among patients and have actively translated the philosophy of patient-centered care and precision medicine into practice. For investigators who seek to conduct translational research, the ATI paradigm serves as a promising example of how to rigorously examine differential effects of intervention on outcomes as a function of individual aptitudes.

References

Caspi, O., & Bell, I. R. (2004). One size does not fit all: Aptitude × treatment interaction (ATI) as a conceptual framework for complementary and alternative medicine outcome research. Part 1—What is ATI research? *The Journal of Alternative and Complementary Medicine,* *10*(3), 580–586.

Chiesa, A., Calati, R., & Serretti, A. (2011). Does mindfulness training improve cognitive abilities? A systematic review of neuropsychological findings. *Clinical psychology review, 31*(3), 449–464.

Collins, F. S., & Varmus, H. (2015). A new initiative on precision medicine. *New England Journal of Medicine, 372*(9), 793–795.

Committee on Quality Health Care in America, Institute of Medicine. (2001). *Crossing the quality chasm: A new health system for the 21st century.* Washington, D.C: National Academy Press.

Cronbach, L. J. (1957). The two disciplines of scientific psychology. *American psychologist, 12*(11), 671.

Epstein, R. M., & Street, R. L. (2011). The values and value of patient-centered care. *Annals of Family Medicine, 9*(2), 100–103.

Gregerson, M. B., Roberts, I. M., & Amiri, M. M. (1996). Absorption and imagery locate immune responses in the body. *Biofeedback and Self-Regulation, 21*(2), 149–165.

Gu, J., Strauss, C., Bond, R., & Cavanagh, K. (2015). How do mindfulness-based cognitive therapy and mindfulness-based stress reduction improve mental health and wellbeing? A systematic review and meta-analysis of mediation studies. *Clinical psychology review, 37,* 1–12.

Karno, M. P., Beutler, L. E., & Harwood, T. M. (2002). Interactions between psychotherapy procedures and patient attributes that predict alcohol treatment effectiveness: A preliminary report. *Addictive Behaviors, 27*(5), 779–797.

MacCoon, D. G., Imel, Z. E., Rosenkranz, M. A., Sheftel, J. G., Weng, H. Y., Sullivan, J. C., . . . Lutz, A. (2012). The validation of an active control intervention for Mindfulness Based Stress Reduction (MBSR). *Behaviour research and therapy, 50*(1), 3–12.

MacCoon, D. G., MacLean, K. A., Davidson, R. J., Saron, C. D., & Lutz, A. (2014). No sustained attention differences in a longitudinal randomized trial comparing mindfulness based stress reduction versus active control. *PloS one, 9*(6), e97551.

McCubbin, T., Dimidjian, S., Kempe, K., Glassey, M. S., Ross, C., & Beck, A. (2014). Mindfulness-based stress reduction in an integrated care delivery system: One-year impacts on patient-centered outcomes and health care utilization. *The Permanente Journal, 18*(4), 4.

Maizes, V., Rakel, D., & Niemiec, C. (2009). Integrative medicine and patient-centered care. *Explore: The Journal of Science and Healing, 5*(5), 277–289.

Mussell, M. P., Mitchell, J. E., Crosby, R. D., Fulkerson, J. A., Hoberman, H. M., & Romano, J. L. (2000). Commitment to treatment goals in prediction of group cognitive–behavioral therapy treatment outcome for women with bulimia nervosa. *Journal of Consulting and Clinical Psychology, 68*(3), 432.

National Research Council. (2011). *Toward precision medicine: Building a knowledge network for biomedical research and a new taxonomy of disease.* National Academies Press.

Neff, D. F., Blanchard, E. B., & Andrasik, F. (1983). The relationship between capacity for absorption and chronic headache patients' response to relaxation and biofeedback treatment. *Biofeedback and Self-Regulation, 8*(1), 177–183.

Rathert, C., Wyrwich, M. D., & Boren, S. A. (2013). Patient-centered care and outcomes: A systematic review of the literature. *Medical Care Research and Review, 70*(4), 351–379.

Rocco, N., Scher, K., Basberg, B., Yalamanchi, S., & Baker-Genaw, K. (2011). Patient-centered plan-of-care tool for improving clinical outcomes. *Quality Management in Healthcare, 20*(2), 89–97.

Rosenkranz, M. A., Davidson, R. J., MacCoon, D. G., Sheridan, J. F., Kalin, N. H., & Lutz, A. (2013). A comparison of mindfulness-based stress reduction and an active control in modulation of neurogenic inflammation. *Brain, behavior, and immunity, 27*, 174–184.

Shoham, V., & Rohrbaugh, M. (1995). *Aptitude × treatment interaction (ATI) research: Sharpening the focus, widening the lens. Research foundations for psychotherapy practice* (pp. 73–95). Wiley.

Snow, R. E. (1991). Aptitude-treatment interaction as a framework for research on individual differences in psychotherapy. *Journal of consulting and clinical psychology, 59*(2), 205.

Steiger, N. J., & Balog, A. (2010). Realizing patient-centered care: Putting patients in the center, not the middle. *Frontiers of Health Services Management, 26*(4), 15–25.

Chapter 2

Personality and meditation

We detect differences among individuals through observation. Except the most obvious differences in physical appearances, we, as social animals, immediately notice and pick up on differences in personality traits presented among people around us. These observations are gradually deepened through further communication and interaction in various settings. It is without a doubt that differences in personality and dispositional traits are the second-most intuitive interindividual variabilities we constantly encounter in everyday life. Hence the study of individual differences has really flourished in personality psychology, which is a subdiscipline of psychology dedicated to the investigation of sets of characteristic patterns in response to external environment and how they influence on human behavior, lifespan development, cognition, and well-being (Pervin & John, 1999). However, the saying "no two people are alike, yet no two people are unlike" rings true. Even for personality traits there are well-defined and established frameworks that summarize our observations of different types of personalities into meaningful and measurable categories. These categories, or more appropriately, dimensions, can help us characterize common personality traits shared among people, although these traits vary in intensity at the individual level. As we emphasized in Chapter 1, Theoretical frameworks of individual differences in interventions, having a theoretical framework that draws generalizations of observed phenomena while also attending to variability among individuals is particularly useful for scientific investigations.

Before delving into personality traits related to psychological interventions, we first provide the background of the term "personality" as well as clarification of terms such as "temperament" and "character" that share semantic meanings with personality. Let us begin with the first sign of personality in humans. From early infancy, some babies get distracted easily and are fearful of the external environment, while others are quite active and energetic, seeking novel stimuli and can get absorbed by something for an extended period of time. These variabilities presented among infants constitute the beginning of personality development and are often referred to as temperament. Temperament has been theorized to have a biological basis, involving not only a genetic contribution, but also neural networks supporting these individual differences in emotional, motor, and attentional

The Neuroscience of Meditation. DOI: https://doi.org/10.1016/B978-0-12-818266-6.00003-4

reactivity to external environments (Rothbart, 2007). Given the biological bases of temperament, researchers have generally agreed that temperament dimensions are fairly stable across development. However, there has not been a clear distinction between temperament and personality. The best way to understand the distinction conceptually is to treat temperament as an innate predisposition to react in a certain manner and personality as more sophisticated and acquired patterns of thoughts and behavior (McCrae et al., 2000). Therefore temperament is not the same as personality, but it is more like a precursor to personality. As for measuring temperaments in people, it is typically explored in infants and children through parental reports using psychometrically validated questionnaires. Researchers have also developed adult temperament questionnaires for assessing temperament across development. Interestingly, though not surprisingly, a considerable amount of association can be found between major dimensions of temperament and personality. For instance, extraversion from the Big Five personality model is related to the extraversion/surgency temperament dimension, neuroticism is related to negative affectivity and conscientiousness is related to effortful control (Evans & Rothbart, 2007). Character is also a manifestation of individual differences, but unlike personality or temperament, character involves a person's moral worth and responsibility, which to some extent are more deliberate (Banicki, 2017). Overall, personality, temperament, and character are related, yet different, sets of constructs of individual differences in behavior and reactivity to external stimuli and environments.

The focus of our discussion is on personality traits and their implications for influencing psychological interventions and, consequently, their associated outcomes. The reason for this focus will become clear as the discussion unfolds, but the most obvious explanation is that there is a wider range of traits and characteristics in personality than in temperament and character, some of which are related to each other, while others are completely unrelated. Take narcissism as an example. One can characterize this trait in personality dimensions, but cannot categorize it into temperament or character, as the former is primarily biologically based constructs and the latter is related to moral worth and responsibility.

Big Five factor model of personality

The most prominent and widely accepted personality framework is the Big Five (or five-factor model) of personality proposed by McCrae and Costa in the 1980s. The five-factor model includes extraversion (or surgency), conscientiousness, neuroticism, agreeableness, and openness to experience, which can be measured by the NEO Five Factor Inventory. However, Eysenck, Barrett, and Eysenck (1985) argue that a three-factor model is sufficient, with extraversion, neuroticism, and psychoticism as the three personality dimensions. Nonetheless, the five-factor model and its well-validated self-report inventory remain the most recognized framework with respect to scientific investigations of personality and individual differences (Pervin & John, 1999). Although the two approaches disagree on the number of dimensions constituting a person's personality, there is consensus on the meaning of each dimension (Digman, 1990).

Extraversion or surgency denotes the tendency to seek out social attention and companionship, exhibit assertive behavior and decisive thinking, and be outgoing and talkative in interpersonal relationships (McCrae & Costa, 2003). People with high levels of extraversion are usually at the center of the spotlight and are often energetic and enthusiastic in most social contexts. There has also been some promising evidence showing that extraversion is predictive of positive affect, suggesting that it may play a role in promoting psychological well-being (Gutiérrez, Jiménez, Hernández, & Pcn, 2005; McCrae & Costa, 2003). Yet, one should always be careful when drawing a direct positive relationship between extraversion and positive affectivity as they are neither equivalent in meaning nor in an empirical sense. Imagining a scenario in which extraverts do not get the social attention or interaction they desire, or they are not given the opportunity to be assertive and outgoing, would they become more upset than people who are low in extraversion and who are also not interested in social interaction and excitement? The answer is not exactly clear without empirical investigations, but this does suggest a possibility that extraversion could also be related to aspects of negative affectivity given different situations. Therefore we must consider contexts and environment when making inferences of empirical relationships among personality dimensions and constructs related to emotion and subjective well-being.

Conscientiousness is a highly examined dimension of personality, referring to the tendency to be organized, diligent, industrious, and achievement-oriented (McCrae & Costa, 2003). Additionally, conscientious people also tend to have a strong sense of responsibility toward life (Pervin & John, 1999). Given these common connotations of conscientiousness, this dimension of personality is often implicated in character since it seems to describe all the "good" qualities one would hope to find in individuals with "good" character (McCrae & Costa, 2003), that is, self-disciplined, thorough, and highly persistent in pursuing goals. There are two aspects of conscientiousness as defined by different theoretical frameworks: (1) regulating behavior and inhibiting unwanted behavior due to conscience; and (2) regulating behavior and inhibiting unwanted behavior due to diligence. We know that conscientious individuals are effective at regulating and inhibiting their behavior, but the motivation behind such regulation can come from either conscience (related to a moral aspect of character) or diligence (having a clear sense of goals). In reality both conscience and diligence seem to covary. With regard to its relation to positive affect, conscientiousness also seems to play a role in contributing to factors that lead to psychological well-being. For example, studies have shown that people with high levels of conscientiousness utilize strategies that are more effective for coping with stress, serving as a potential protective factor against daily pressure and contributing to aspects of positive affect (Bartley & Roesch, 2011; McCrae & Costa, 2003).

Neuroticism can be best described as the tendency to experience negative affects including, but not limited to, emotional distress, frustration, anger, and sadness as well as the cognitive and behavioral reactions following these negative experiences (McCrae & John, 1992). It is in some sense the exact opposite of the definition of psychological well-being and mental health, as neuroticism is also referred to as negative emotionality or negative affectivity. We all know that emotional ups and down are common in our everyday life, but the degree to which such emotional experiences affect our daily functioning are very different among people, especially for those who are high in neuroticism where the intensity and frequency of negative affect is out of proportion to the challenges they encounter (McCrae & Costa, 2003). For instance, they may engage in self-criticism, ruminate on the upsetting experiences, and can't seem to cope with negative events. Not surprisingly, high levels of neuroticism is associated with chronic negative effects, often leading to a wide range of psychiatric disorders, physical tensions, and health-related problems (Lahey, 2009). However, this is not to say that people with low levels of neuroticism are absolutely free from emotion-related and health-related problems or high in positive affect and mental health. A low level of neuroticism may simply imply that these individuals are not highly reactive to negative experiences and may be calmer and more relaxed. The predictive power of neuroticism in mental disorders such as depression is particularly strong when considered in the context of negative life events, as demonstrated by multiple large-scale studies (Lahey, 2009). Furthermore, unlike the other two dimensions, neuroticism has also been included in all major theoretical frameworks and models of personality and is heavily implicated in domains of mental health.

Agreeableness is another personality dimension relevant to character and refers to the tendency to be altruistic, warm, friendly, and cooperative toward other individuals as well as in interpersonal relationships (Pervin & John, 1999). Agreeableness encompasses qualities that would be labeled as "good" character, and those who are more agreeable, are also more supportive of others and are more compliant in most social situations (McCrae & Costa, 2003). We expect individuals with a high level of agreeableness to maintain a positive and constructive relationship with others and have a higher level of motivation to engage in prosocial behaviors that would facilitate their relationship goals (Graziano & Tobin, 2009). More importantly, people who are more agreeable report higher levels of subjective well-being, life satisfaction, and emotional thriving, suggesting agreeableness, along with extraversion and conscientiousness, may together contribute to psychological health (Friedman & Kern, 2014).

Openness to experience refers to a sense of curiosity, open-mindedness, and acceptance of novel experiences (McCrae & Costa, 2003). Individuals with high levels of openness to experiences may engage in unconventional activities, are often not afraid to try out new experiences, and are also very creative since they tend not to reject novel ideas and information. However, given that these highly open individuals exhibit high creativity and capacity of divergent thinking, early discussion of its definition was rather controversial. It is perhaps the most controversial personality dimension out of the five dimensions due to some theorists considering intellect as a part of this dimension. However, it quickly became evident that even though people who are high in openness to experience and have high cognitive abilities with respect to creativity and divergent thinking, the empirical association between openness to experience and intellect (or intelligence) was relatively weak (McCrae & Costa, 1985). Similar to other dimensions of personality (except neuroticism), studies have shown that a high level of openness to experience is positively related to psychological well-being (Lamers, Westerhof, Kovács, & Bohlmeijer, 2012). This relationship is intuitive because individuals with high levels of openness to experience are more likely to change their maladaptive behaviors and habits when needed.

In summary, providing a thorough and comprehensive description of all five personality dimensions is not possible given the limited scope of this chapter. We briefly discussed each of the five dimensions of personality based on the most notable five-factor model, hoping to provide a basic understanding of these major dimensions to facilitate our discussion of how these personality traits are related to psychological interventions and their effectiveness in improving the targeted outcomes.

Big Five factor and intervention effectiveness

The impact of personality on human behavior is obvious. Based on our discussion of conceptual frameworks in Chapter 1, Theoretical frameworks of individual differences in interventions, personality traits have promising potential to serve as individual differences variables that interact with psychological interventions to influence intervention outcomes. Indeed, a growing body of research evidence has already indicated that individual differences in personality traits are associated with differential intervention effectiveness (Chapman, Hampson, & Clarkin, 2014; Gully & Chen, 2010). In particular, the Big Five dimensions of personality have long been theoretically hypothesized and empirically shown to influence the extent to which individuals benefit from psychological interventions, including contemplative practices such as meditation.

We have described extraversion as the tendency to be talkative and assertive in interpersonal interactions and relationships and have also suggested that it may be related to aspects of psychological well-being. When we consider extraversion within the context of psychological interventions, we would immediately realize that extraversion may play a role in group-based interventions where interpersonal and social interactions are unavoidable. There is evidence suggesting that people with high levels of extraversion tend to prefer interventions involving interpersonal interactions rather than a pure lecture-style intervention program where instructors are the primary speaker (Anderson, 1998; Sanderson & Clarkin, 1994). Conversely, individuals with low extraversion (more introverted) tend to prefer a more structured and goal-directed intervention approach than highly interactive group-based intervention (Anderson, 1998; Sanderson & Clarkin, 1994). Based on our experience and research findings, we know that preferences for specific intervention approaches have practical implications for intervention effectiveness, since individuals would likely be more engaged and active in their preferred interventions.

For instance, one study examined the effects of two types of intervention among women with histories of childhood sexual abuse and assessed individual differences in baseline personality traits. The study found that for women who are low in extraversion and agreeableness, better intervention outcomes can be observed if the intervention was provided in a highly structured and skill-focused manner, which in this particular study was Women's Safety in Recovery (WSIR) that focuses on problem-solving exercises (Talbot, Duberstein, Butzel, Cox, & Giles, 2003). In contrast, for the common treatment as usual approach that is largely devoted to group therapy to discuss general strategies of crisis resolution and symptom reduction, women with low levels of extraversion and agreeableness did not show the same degree of improvement. Although research has yet to examine the potential influence of extraversion on meditation-based interventions, we can infer from these results that extraverts are likely to enjoy the discussion aspect of typical meditation-based interventions in which people usually go around a circle to share their experiences and raise questions. However, there is also the practice-heavy aspect of meditation-based intervention, in which people are asked to meditate silently. One of the meditation-based interventions, the mindfulness-based stress reduction program, includes a one-day silent retreat that asks everyone to practice silently for several hours. It is unlikely that individuals with high levels of extraversion would enjoy the silent retreat. Nonetheless, without specifically investigating extraversion as a potential moderator of meditation effects, it is difficult to conclude whether or not it would have a substantial impact on the targeted outcomes.

Conscientiousness seems to be one of the most desirable personality traits we hope to find in most individuals. It represents qualities such as prudence, persistence, organization, industriousness, and dependability, which are all valuable for achieving life goals and maintaining health throughout the lifespan (Friedman & Kern, 2014). It has also been found to be highly predictive of health and longevity across development, suggesting its critical role in promoting psychological well-being and perhaps even a causal role in establishing a healthy lifestyle, important for ensuring physical and mental well-being. With respect to meditation practices, one possibility may be that individuals with high levels of conscientiousness could be drawn to contemplative practices such as meditation that is beneficial for promoting health and well-being. If this hypothesis is true, then highly conscientious individuals are motivated to learn and practice meditation, thereby receiving the most positive effects out of meditation-based interventions. Emerging evidence suggests that highly conscientious individuals are more likely to engage in healthier behaviors (Friedman & Kern, 2014) and meditation is undoubtedly one of the most popular lifestyle interventions a person could participate in in recent years. Conscientious individuals tend to get acquainted with other conscientious individuals who share the same values and goals, which in some sense is another social reinforcer that propel people with high levels of conscientiousness into sustaining their healthy behaviors and lifestyle. The same is not true for individuals with low levels of conscientiousness as they are likely to experience more detrimental health consequences due to a wide range of physical, emotional, and cognitive problems. However, this does not necessarily mean that people with low conscientiousness would be unsuitable for psychological interventions. They may not be drawn to meditation voluntarily, but interventions that promote psychological health and cognitive function should, in theory, be of benefit to them. Although the direct evidence for the role of conscientiousness in meditation-based interventions is not available, increasing research attention should be paid toward this promising dimension of personality.

Another interesting aspect of conscientiousness is its association with intervention adherence and compliance (Chapman et al., 2014; Sanderson & Clarkin, 1994). According to a meta-analysis of conscientiousness, this personality dimension is positively related to motivation to learn, which is also a factor highly correlated with intervention effectiveness (Colquitt, LePine, & Noe, 2000). We have touched upon the idea of motivation from a conceptual standpoint. Here, direct empirical evidence seems to suggest that motivation may be one of the underlying factors contributing to adherence to intervention instructions and protocols. In particular, for highly conscientious individuals, this motivation and adherence may ultimately manifest as improvement in intervention outcomes. For instance, a study on the effects of cognitive training on improving working memory found that individuals with high conscientiousness showed more improvement in working memory performance after training completion and also reported high training enjoyment (Studer-Luethi, Jaeggi, Buschkuehl, & Perrig, 2012). These results corroborate the notion that conscientiousness is closely associated with intervention effectiveness through potentially affecting the motivation to learn, which, in turn, influences the degree of intervention enjoyment as well as the commitment to intervention goals (Komarraju & Karau, 2005). We can generalize this putative underlying mechanism of how conscientiousness indirectly affects intervention effectiveness to meditation-based interventions, since motivation and commitment are essential to any intervention program. However, it is still important to realize that conscientiousness may also interact with other mental processes such as persistence and industriousness that would likely facilitate the acquisition and maintenance of specific skills and practices.

Moving onto neuroticism, the tendency to experience negative affect and emotional instability has generally been found to make individuals nonresponsive or less responsive to most types of psychological intervention (Anderson, 1998). This is not surprising as neuroticism is linked with various mental disorders (e.g., depression, eating disorders, and schizophrenia) and symptoms, as well as physical health problems not included in the neuroticism items (Lahey, 2009). These deficits in emotion and health undoubtedly influence the daily functioning of individuals, impeding them from gaining benefits out of psychological interventions that typically require attention and aspects of cognition in order to learn and practice. However, this is not to say that individuals with high levels of neuroticism are hopeless, instead this low responsiveness to psychological interventions seems to suggest that alternative intervention approaches may be more effective. For instance, one study demonstrated that in patients with major depressive disorder (MDD), high neuroticism is associated with more improvement in depressive symptoms at postintervention if pharmacotherapy (PHT) was administered (Bagby, Quilty, Segal, McBride, Kennedy, & Costa, 2008). The alternative intervention in this experiment was cognitive behavior therapy (CBT), a close analog to meditation-based interventions. Consistent with what has been found before, individuals with high levels of neuroticism did show less improvement in depressive symptoms when psychological intervention (CBT) was used rather than PHT. It is again worth noting that when intervention effects on depressive severity were averaged within groups, both intervention approaches (PHT and CBT) significantly reduced depressive symptoms, and there were no group differences between CBT and PHT in terms of intervention effectiveness. However, it is obvious that differential effects on depressive symptoms can be observed across individuals if differences in personality are considered. This highlights the importance of examining individual variability in psychological intervention research since the standard group-average approach neglects critical information that is imperative for clinical application.

With regard to meditation-based interventions, people with high levels of neuroticism are not necessarily going to be nonresponsive. In fact, one study of mindfulness-based stress reduction program in healthy populations showed that individuals with higher levels of neuroticism exhibited greater improvement in subjective well-being and psychological distress (de Vibe et al., 2015). Likewise, in a separate study, healthy individuals with higher levels of neuroticism also showed greater reduction in symptoms of depression and anxiety immediately after meditation-based intervention and during a 3-month follow-up (Nyklíček & Irrmischer, 2017). Thus the critical consideration is then which intervention we should provide to individuals given their different levels of neuroticism, their clinical diagnoses, and our targeted outcomes. From existing research evidence, it seems that for people with subclinical levels of anxiety and depression and high levels of neuroticism, meditation-based interventions would be effective in achieving the desirable outcomes related to mental health and psychological well-being. On the other hand, for those who are high in neuroticism and clinically diagnosed with depression and other related disorders, engaging in meditation-based interventions may be too demanding given the suboptimal cognitive and emotional states of these individuals, which makes taking medication the most effective approach for treating psychological symptoms. As for the cognitive outcomes of meditation, one study has shown that the interaction of mood states and neuroticism is related to improvement in creativity following meditation-based intervention (Integrative Body-Mind Training), such that individuals with lower neuroticism exhibited greater improvement when they had higher vigor in mood state (Ding, Tang, Deng, Tang, & Posner, 2015). This study offered a different perspective on the role of neuroticism, that is, lower neuroticism was associated with greater improvement in cognitive outcome, which is contrary to what has been found with regard to other psychological outcomes.

Without further validation of these finding, it is difficult to know whether neuroticism has a consistent impact on meditation-based interventions, or what the exact cutoff point for low/average versus high neuroticism in both healthy and clinical populations would be. For instance, one review proposed using neuroticism as a screening index to identify individuals with high risk of mental health problems due to high neuroticism. The idea is to then provide these high-risk individuals with preventive intervention that would hopefully reduce the likelihood of adverse outcomes (Lahey, 2009). However, without a benchmark from which to screen vulnerable individuals, we cannot really utilize this knowledge in practice. More studies are needed to determine whether higher neuroticism can also lead to better outcomes in other domains of individual well-being such as cognitive function and physical health. Overall, the moderating effect of neuroticism on intervention responsiveness suggests the utility of this personality dimension in predicting the outcomes of psychological interventions and may be particularly promising for clinical applications.

Agreeableness can be operationalized as friendliness, warmth, and cooperativeness within social interaction and interpersonal relationships. We have discussed a study on the role of agreeableness and extraversion in influencing intervention effectiveness in women with childhood abuse histories. The study found that agreeableness also has the potential to moderate intervention effects on targeted outcomes, such that highly agreeable individuals, just like extraversion, seem to respond better to group-based interventions that engage interpersonally oriented strategies, whereas less agreeable individuals would likely find this approach less helpful and somewhat confrontational (Talbot et al., 2003). These results again suggest that understanding individual differences in personality dimensions is highly informative for choosing the most appropriate type of interventions for each individual. In theory, providing interventions that satisfy individual preferences should promote intervention adherence and engagement (Christensen & Johnson, 2002), and increase individuals' satisfaction as described by research on patient-centered care. Consequently these processes could together increase the effectiveness of intervention in improving targeted outcomes. Unfortunately for meditation-based interventions that mix both interpersonal interactions and individual practices, we are still lacking adequate evidence to determine its exact role in affecting intervention outcomes. Aside from the influence of agreeableness in preferences for specific intervention approaches, this interesting personality dimension is also related to attrition in experimental studies. For instance, the likelihood of someone completing a research study can vary across people and it has been shown that individuals with high levels of agreeableness (as well as being highly conscientious) are less likely to withdraw from ongoing studies, suggesting that they may be more compliant and tend to adhere to instructions and what is expected of them (Friedman & Kern, 2014). Given that agreeableness and conscientiousness are involved, it is difficult to separate how much each dimension contributes to low attrition. Therefore it is possible that these two personality dimensions potentially work together to enhance intervention adherence and compliance. However, there is also another possibility that agreeable individuals may generally benefit more from interventions due to their compliant and easygoing attitude, and that preferences for interpersonal approaches may only be one of the underlying factors contributing to differential intervention effectiveness. Taken together, the possibilities and hypotheses still necessitate rigorous scientific investigations for validation and confirmation. For meditation-based interventions, the implications of agreeableness in affecting intervention effectiveness and outcomes are highly relevant for future studies and clinical applications.

Openness to experience represents an overall curiosity and acceptance of new experiences and ideas and has been linked to creativity and divergent thinking. Interestingly, among a wide variety of possible experiences, individuals with high levels of openness are more likely to be attracted to complementary and alternative medicine (CAM) techniques such as yoga and meditation (Thomson, Jones, Browne, & Leslie, 2014). They are also more likely to use and practice these CAM techniques in daily life, which could suggest that these individuals may have better psychological health and wellbeing. If we were to put those with an affinity for CAM techniques into meditation-based intervention, maintaining regular meditation practices outside of group-based intervention contexts would not be a problem. This regular practice could potentially lead to better intervention outcomes at postintervention. A different study examined whether personality traits could influence the frequency of meditation practices and found that individuals with high levels of openness to experience engaged in more mindfulness practices during and after an 8-week, meditation-based, intervention program (Barkan et al., 2016), further supporting the observed trend from the prior study. More importantly the tendency to engage in more practice was still true even when demographic differences in age, educational level, and sex were controlled for, suggesting the unique impact of openness to experience on meditation practice behavior. Although neither of these studies directly investigated the relationship between practice time and intervention outcomes, previous research in meditation-based intervention has suggested a positive relationship between practice time and improvement in outcomes such as cognitive function and brain functional and structural plasticity (Brewer et al., 2011; Chan & Woollacott, 2007; Jha, Stanley, Kiyonaga, Wong, & Gelfand, 2010). A separate line of evidence showed that individuals with high levels of openness to experience also benefited more from intervention approaches devoted to self-exploration and discovery, which are largely similar to most meditation-based techniques found in typical intervention programs that heavily emphasize self-awareness and introspection (Anderson, 1998). Openness to experience seems to have a special role in moderating the effects of meditation-based interventions.

Based on these findings we can speculate that the key mechanism through which openness to experience influences meditation-based intervention effects is increasing the amount of practice time and the frequency of practice. However, the question is why individuals with high openness to experience are attracted to contemplative practices in the first place. One of the reasons for this is that when people with high openness to experience encounter health-related problems they tend to seek out nonconventional care approaches. Notably, most individuals use CAM techniques in conjunction with their conventional medical care services and do not first seek out CAM approaches as their primary treatment (Thomson et al., 2014). We know that being openminded often entails less adherence to conventional thinking and beliefs, suggesting that individuals with high openness to experience are in general more likely to engage in alternative strategies (i.e., CAM) for improving their health and well-being, even without the presence of health problems. As for practical implications, we know that within treatment contexts, respecting individual preferences and choices are important, but there is also a downside to this approach in that a more effective intervention, other than the one the patient chooses, could have already existed. For instance, what if taking medication is more effective in achieving the targeted outcomes than meditation-based interventions for people with high openness to experience, would we still want to stick with meditation? Unfortunately, we do not have enough knowledge on the role of openness to experience in other types of intervention or treatment approaches; thus, it is impossible to compare the effectiveness of interventions in relation to individual differences in openness to experience. Future studies would need to address this research gap in order to better elucidate the role of openness to experience in affecting intervention effectiveness.

In this section, we discussed how each personality dimension may moderate the effects of psychological interventions on outcomes associated with mental health and individual well-being. We specifically focused on the relevance of all five dimensions for meditation-based interventions and indicated that each personality dimension may influence intervention outcomes through distinct mechanisms. We also suggested the possibility that these dimensions may interact with each other to moderate intervention effectiveness. In light of existing empirical evidence, future investigations would need to conduct studies directed toward delineating the underlying mechanisms through which personality dimensions influence meditation-based intervention effectiveness and outcomes. Future research would also need to examine the influence of personality in intervention effectiveness using a comparative study design in which meditation-based intervention is compared against another alternative intervention. This would allow us to address one important practical question, that is, which intervention would work best for whom, and for whom such intervention benefits would be most pronounced. This question lies at the center of our discussion throughout the book since, fundamentally, we would want to be able to utilize our knowledge of individual differences from scientific investigations to inform our clinical applications and practices in a meaningful way.

Trait mindfulness

Trait mindfulness (or sometimes called dispositional mindfulness) is perhaps the most relevant personality trait to date for meditation-based interventions. It refers to the innate capacity of paying and maintaining attention to present-moment experiences with an open and nonjudgmental attitude (Brown & Ryan, 2003). This definition of trait mindfulness may sound familiar to many readers as it is also one of the key goals of meditation practices, that is, to promote present moment attention and awareness. Trait mindfulness is typically measured using self-report questionnaires, although there are at least 10 different questionnaires available for assessing this particular trait. Currently there is no real consensus on which questionnaire can be considered to be the most valid and reliable instrument, thus we would like to point out two of the most popular questionnaires that have been widely used in literature: (1) mindful attention awareness scale (MAAS) (Brown & Ryan, 2003); and (2) five facet mindfulness questionnaire (FFMQ) (Baer, Smith, Hopkins, Krietemeyer, & Toney, 2006). The MAAS only has one summary score representing the level of trait mindfulness whereas FFMQ not only has one summary trait mindfulness score, but also includes five facets of mindfulness, which are: (1) observing—the ability to observe present-moment experiences, (2) describing—the ability to describe present-moment experiences, (3) acting with awareness, (4) nonjudging—the ability to be nonjudgmental of present-moment experiences, and (5) non-reactivity—the ability to be nonreactive to present-moment experiences. It is obvious that FFMQ taps into specific components of trait mindfulness, which potentially could be more accurate in separating components that are specifically related to other measures of psychological well-being, rather than the overall mindfulness capacity. Regardless of which questionnaire we use, prior research has established an overall positive relationship between trait mindfulness and psychological well-being (Brown & Ryan, 2003; Keng, Smoski, & Robins, 2011), suggesting that trait mindfulness could be one of the indicators of psychological health and may be a critical factor contributing to overall psychological health. This interesting hypothesis is further supported by studies showing that mind-wandering (the tendency of the mind to wander away from the present moment) is associated with negative effects and psychological distress (Killingsworth & Gilbert, 2010; Stawarczyk, Majerus, Van der Linden, & D'Argembeau, 2012). Furthermore, not only is mind-wandering problematic for psychological well-being, but also entails the general tendency to pay less attention and awareness to what one is currently doing. It is important to distinguish mind-wandering from low levels of trait mindfulness, as these are related constructs but not necessarily the same.

Given the conceptual relevance of trait mindfulness for meditation practices, it is not difficult to imagine that a baseline level of trait mindfulness would at least have some impact on the effectiveness of meditation-based interventions, especially for individuals who first start to practice meditation. One hypothesis is that individuals with higher trait mindfulness would likely have an easier time learning and practicing meditation than those who tend to get distracted by things that are not happening at the present moment. One interesting theory regarding trait mindfulness proposed that meditation-based interventions or training programs which repeatedly induce mindfulness states may lead to more stable and trait-level changes in trait mindfulness (Kiken, Garland, Bluth, Palsson, & Gaylord, 2015). Not surprisingly, in a study of an 8-week, meditation-based intervention, individuals with high levels of trait mindfulness at baseline showed more improvement in trait mindfulness at postintervention (Shapiro, Brown, Thoresen, & Plante, 2011). Although the result is preliminary and there is missing evidence on the influence of trait mindfulness in other psychological outcomes of meditation, the positive association between trait mindfulness and psychological well-being implies that the observed differential effects may also apply to other measures of psychological health.

For cognitive outcomes of meditation-based interventions, trait mindfulness also has the potential to influence the extent of improvement in these outcomes. Research has consistently demonstrated positive correlations between trait mindfulness and task performance in different cognitive control tasks (Ruocco & Wonders, 2013; Schmertz, Anderson, & Robins, 2009). However, for cognitive functions, not only is the general capacity of mindfulness related to cognitive function, but different facets of trait mindfulness are also uniquely related to specific aspects of cognitive function. For instance, one study showed that individuals with high levels of observing had better visual working-memory performance, whereas those high in non-reactivity showed greater cognitive control flexibility (Anicha, Ode, Moeller, & Robinson, 2012). Although we still do not have direct evidence of trait mindfulness moderating the effects of meditation-based interventions, these positive associations seem to suggest that trait mindfulness could potentially exert an influence over the effects of intervention on cognitive function, such that those with high trait mindfulness may benefit more from the intervention. Similarly there is also some evidence illustrating that trait mindfulness may potentially mediate the effects of meditation-based intervention on cognitive function through simultaneously attenuating emotional interference and enhancing attentional control (Ortner, Kilner, & Zelazo, 2007; Shahar, Britton, Sbarra, Figueredo, & Bootzin, 2010).

Despite research supporting trait mindfulness as one of the most important contributors for differential intervention effects are mostly preliminary and indirect, trait mindfulness is undoubtedly highly associated with measures of psychological health and cognitive function. Different facets of trait mindfulness have been shown to support specific cognitive abilities, suggesting that trait mindfulness could play a vital role in the underlying mechanisms and processes supporting meditation-related improvement in cognition and mental health. However, we need to recognize that we are at the early stage of individual differences research for psychological interventions, making conclusions about the directionality of the effects of personality traits, including trait mindfulness, on different intervention outcomes are still very premature. Knowing the implications of personality traits for intervention effectiveness and their relationships with different outcome measures of intervention is more important for planning future studies that interrogating the moderating and mediating effects of individual differences variables. Moreover, growing evidence has indicated that meditation practice induces state and trait changes. State change refers to temporary changes of the brain and corresponding patterns of activity or connectivity, whereas trait change alters personality traits following a longer period of practice. It had been traditionally assumed that personality traits are relatively stable entities, but more further research demonstrates that personality can change over time as a result of life experiences or through practice (Nyklíček, van Beugen, & Denollet, 2013; Shapiro et al., 2011; Tang, Holzel, & Posner, 2015), suggesting that personality is itself flexible. Although this demonstrates that individuals can change the way that they feel, believe, and act, it also complicates the systematic investigation of the construct of "trait mindfulness" (Wheeler, Arnkoff, & Glass, 2016). This evidence suggests that it will be important to assess trait mindfulness at different points for studies investigating the effects of meditation. It should also be noted that trait mindfulness is assessed through self-report questionnaires. The use of these questionnaires comes with specific challenges and limitations, which have been critically discussed (Grossman, 2011; Park, Reilly-Spong, & Gross, 2013). It is therefore important to remember that what is interpreted as "trait mindfulness" is what these questionnaires assess.

Trait mindfulness and the Big Five—mindful personality?

Although we discussed trait mindfulness and the five personality dimensions in separate sections, the relationship between these two sets of personality traits deserve closer attention. Is trait mindfulness related to the five personality dimensions? If so, what does it say about trait mindfulness with respect to its implications for intervention effectiveness and outcomes? Is there a mindful personality that can be conceptualized based on these relationships with the five personality dimensions?

A recent meta-analysis on the relationship between trait mindfulness and the five-factor model of personality attempted to replicate and resolve any inconsistences from previous studies focusing on the same relationship. It not only aggregated all studies that examined this relationship in analysis, but also separated out studies that used MAAS from studies that used FFMQ and divided them into two different analyses. This approach is very informative in that if there are any differences in the relationships between the two questionnaires they would allow us to compare and contrast which questionnaire is more sensitive in capturing the relationships, and whether or not these two questionnaires would differ drastically even if they are intended to measure the same construct.

In the meta-analysis, three different sets of analyses were conducted. The first analysis focused on all existing studies of trait mindfulness and the five personality dimensions. As we would have expected (based on what we know about the five-factor model of personality) neuroticism showed a strong negative relationship with trait mindfulness (Hanley & Garland, 2017), conscientiousness exhibited the strongest positive relationship with trait mindfulness followed by a moderate positive correlation between trait mindfulness and agreeableness and very modest positive correlations between trait mindfulness and openness to experience and extraversion. Looking only at studies of trait mindfulness as measured by MAAS, the same trend was observed except that openness to experience and extraversion had even more modest positive correlations with trait mindfulness. For studies that used FFMQ as the measure of trait mindfulness, the overall composite score (the sum of all five facets scores) showed a stronger negative correlation with neuroticism than that of MAAS, as well as stronger positive relationships with extraversion and openness to experience. However, similar magnitudes of correlations were found for agreeableness and consciousness. These results suggest that even though MAAS is a unidimensional measure of trait mindfulness, its relationships with the five personality dimensions are still very similar to that of FFMQ, which can be considered as a multidimensional measure of trait mindfulness with the composite score being the sum of all five facets. The meta-analysis also examined how each facet of FFQM is related to the five personality dimensions. Both extraversion and agreeableness showed similar positive correlations across all five facets, but were most strongly associated with the describing facet (Hanley & Garland, 2017). Neuroticism was associated with four facets, but was most strongly and negatively associated with acting with awareness, nonreacting, and nonjudging. Conscientiousness was also associated with four facets, but was strongly positively correlated with the acting with awareness facet. Last, openness to experience showed the strongest correlation with the observing facet, but was only associated with three of the five facets.

Based on these results, trait mindfulness is robustly and consistently related to neuroticism and conscientiousness regardless of which questionnaire is used to assess trait mindfulness. While extraversion and agreeableness are moderately correlated with the construct, openness to experience showed the lowest magnitude of correlations. Given the positive relationship between trait mindfulness and psychological well-being, and conscientiousness and psychological well-being, it is theoretically understandable to find trait mindfulness correlated with conscientiousness. The same is true for neuroticism since the negative relationship further confirms the notion that trait mindfulness is associated with psychological well-being and not with negative effects or psychological symptoms. If we were to integrate the five-factor model of personality and trait mindfulness into a tentative definition of mindful personality, it is safe to say that a mindful personality refers to a sense of emotional stability and highly organized and disciplined capacity of self-control (Hanley & Garland, 2017). The facet analysis also highlights that not all aspects of trait mindfulness are related to each personality dimension, suggesting that future investigations using either trait mindfulness or personality dimensions as individual differences factors to predict intervention outcomes may want to consider selecting the most robust indexes out of each construct. For instance, conscientiousness and neuroticism may likely have better predictive power, and acting with awareness, nonreacting, and nonjudging may have greater likelihood of showing an effect on intervention outcomes.

Motivation and goals

In addition to personality traits, individual differences in personal goals and motivation could also serve as promising moderators or predictors of intervention outcomes. Goals and motivation are relatively malleable individual characteristics that may vary as a function of different contexts and situations. Personal goals have often been investigated as goal orientation within knowledge-based or skill-based training contexts, which are somewhat different from typical psychological intervention context in that they each target different kinds of changes in behavior (i.e., learning how to perform a task vs. reducing depressive symptoms). Nonetheless, goal orientation has implications for the effectiveness of psychological intervention since participants receiving intervention would inevitably acquire some skills and knowledge associated with targeted behavioral changes. Two distinct types of goal orientation have been defined, namely learning goal orientation (LGO) and performance goal orientation (PGO). The former describes a tendency to increase and develop competence and mastery in specific skills and knowledge taught in training, while the latter solely emphasizes improving performance and exceeding normative standards in specific training outcomes without paying much attention to the learning process itself (Gully & Chen, 2010).

Studies have shown that individuals with high LGO frequently elicit self-regulatory processes to facilitate a deeper comprehension of concepts and skills, leading to positive effects on knowledge coherence (Kozlowski et al., 2001). Furthermore, such individuals also exhibit high motivation to learn and high self-efficacy, which, in turn, positively influence training outcomes (Gully & Chen, 2010). However, high PGO does not result in similar benefits in training outcomes given its positive correlation with outcomes-related interference, such as anxiety arising from the avoidance of failure in performance (Gully & Chen, 2010). Although goal orientation has received less attention within psychological intervention, the findings from training programs in industrial and organizational psychology have pivotal implications for meditation-based intervention. For instance, LGO individuals may focus solely on learning and practicing meditation techniques for personal growth, which could naturally lead to improvement in certain outcomes, while PGO individuals may strive for improvement in specific outcomes through practicing meditation, but inadvertently distract themselves with the anxiety toward potential failure or with validation-seeking behavior by trying to prove that the intervention is working, ultimately resulting in unchanging or worse intervention outcomes as observed in previous studies of knowledge-based training (Gully & Chen, 2010). The role of goal orientation in affecting outcomes across different interventions should not be underestimated since individuals who participate in any form of training or intervention would at least have some goal orientation toward one way or the other, or a mixture of both with varying extent in each.

Looking further into the underlying process that supports the formation of goal orientation and other various behaviors in our lives, motivation stands out as the primary contributor. Motivation to learn plays a critical role in predicting intervention effectiveness and is also closely associated with personality dimensions including extraversion, conscientiousness, and openness to experience. The motivation to change and to discontinue psychologically dysfunctional behavior has been shown to positively predict improvement in behavioral outcomes following psychological intervention (Mussell et al., 2000). Despite the consensus that motivation generally plays a facilitated role in psychological intervention, it is necessary to elucidate and understand the specific type of motivation contributing to differential intervention effectiveness among individuals.

One type of motivation is particularly important within the context of psychotherapy and other related types of intervention, namely autonomous motivation. Autonomous motivation refers to the extent to which individuals experience the freedom of choice and setting goals that do not involve external pressure (incentives or punishment) or internal guilt that may prompt them into certain decisions (Deci & Ryan, 2000). This concept is the core of self-determination theory, which distinguishes this form of motivation from controlled motivation such as performing an action in order to receive reward or avoid punishment (Zuroff et al., 2007). In a methodologically rigorous examination, Zuroff et al. investigated the relationship between autonomous motivation and intervention outcomes in depressed patients who underwent one of three interventions, namely CBT, interpersonal therapy, or pharmacotherapy with clinical management. Results replicated findings of previous empirical reports, showing that autonomous motivation significantly predicted better outcomes. In particular, autonomous motivation predicted higher probability of achieving remission and lower depression severity across all three interventions (Zuroff et al., 2007). These findings not only illustrate the catalyzing effects of autonomous motivation across different psychological interventions, but also suggests individual differences in such indexes before the onset of intervention can strongly influence the extent of improvement in final outcomes.

Summary and implications for translational research

In this chapter we extensively discussed how personality traits and individual differences in motivation and goals may moderate and predict intervention effects on psychological and cognitive outcomes. Although each personality trait seems to differentially relate to intervention outcomes, the intercorrelations among personality traits and characteristics are evident throughout our discussion. For example, personality traits are associated with motivation and may exert its influence over intervention outcomes through affecting motivation. Likewise, trait mindfulness, the most relevant individual differences variable for meditation-based intervention, is associated with a wide range of psychological outcomes and personality dimensions even without intervention. These interesting relationships with personality dimensions may serve as underlying mechanisms of differential intervention effectiveness. Together these individual differences variables create challenges for future research in that it may be difficult to isolate unique contributions from each variable given their relationships, but they also provide exciting research opportunities to delineate the intricate interaction among these variables and investigate how they influence intervention effectiveness. From a translational perspective, utilizing individual differences in personality traits and related characteristics to make predictions about which intervention would be the most effective and appropriate for which individuals are priorities in contemporary medicine as this would greatly enhance the effectiveness and cost-effectiveness of intervention and treatment in an unprecedented way.

References

Anderson, K. W. (1998). Utility of the five-factor model of personality in psychotherapy aptitude-treatment interaction research. *Psychotherapy Research*, 8(1), 54−70.

Anicha, C. L., Ode, S., Moeller, S. K., & Robinson, M. D. (2012). Toward a cognitive view of trait mindfulness: Distinct cognitive skills predict its observing and nonreactivity facets. *Journal of Personality*, 80(2), 255−285.

Baer, R. A., Smith, G. T., Hopkins, J., Krietemeyer, J., & Toney, L. (2006). Using self-report assessment methods to explore facets of mindfulness. *Assessment*, 13(1), 27−45.

Banicki, K. (2017). The character−personality distinction: An historical, conceptual, and functional investigation. *Theory & Psychology*, 27(1), 50−68.

Barkan, T., Hoerger, M., Gallegos, A. M., Turiano, N. A., Duberstein, P. R., & Moynihan, J. A. (2016). Personality predicts utilization of mindfulness-based stress reduction during and post-intervention in a community sample of older adults. *The Journal of Alternative and Complementary Medicine*, 22(5), 390−395.

Bartley, C. E., & Roesch, S. C. (2011). Coping with daily stress: The role of conscientiousness. *Personality and Individual Differences*, 50(1), 79−83.

Bagby, R. M., Quilty, L. C., Segal, Z. V., McBride, C. C., Kennedy, S. H., & Costa, P. T., Jr (2008). Personality and differential treatment response in major depression: a randomized controlled trial comparing cognitive-behavioural therapy and pharmacotherapy. *The Canadian Journal of Psychiatry*, 53(6), 361−370.

Brewer, J. A., Worhunsky, P. D., Gray, J. R., Tang, Y. Y., Weber, J., & Kober, H. (2011). Meditation experience is associated with differences in default mode network activity and connectivity. *Proceedings of the National Academy of Sciences of the United States of America*, 108(50), 20254−20259.

Brown, K. W., & Ryan, R. M. (2003). The benefits of being present: Mindfulness and its role in psychological well-being. *Journal of Personality and Social Psychology, 84*(4), 822.

Chan, D., & Woollacott, M. (2007). Effects of level of meditation experience on attentional focus: Is the efficiency of executive or orientation networks improved? *The Journal of Alternative and Complementary Medicine, 13*(6), 651–658.

Christensen, A. J., & Johnson, J. A. (2002). Patient adherence with medical treatment regimens: An interactive approach. *Current Directions in Psychological Science, 11*(3), 94–97.

Chapman, B. P., Hampson, S., & Clarkin, J. (2014). Personality-informed interventions for healthy aging: Conclusions from a National Institute on Aging work group. *Developmental psychology, 50*(5), 1426.

Colquitt, J. A., LePine, J. A., & Noe, R. A. (2000). Toward an integrative theory of training motivation: a meta-analytic path analysis of 20 years of research. *Journal of applied psychology, 85*(5), 678.

de Vibe, M., Solhaug, I., Tyssen, R., Friborg, O., Rosenvinge, J. H., Sørlie, T., ... Bjørndal, A. (2015). Does personality moderate the effects of mindfulness training for medical and psychology students? *Mindfulness, 6*(2), 281–289.

Deci, E. L., & Ryan, R. M. (2000). The "what" and "why" of goal pursuits: Human needs and the self-determination of behavior. *Psychological Inquiry, 11*(4), 227–268.

Digman, J. M. (1990). Personality structure: Emergence of the five-factor model. *Annual Review of Psychology, 41*(1), 417–440.

Ding, X., Tang, Y. Y., Deng, Y., Tang, R., & Posner, M. I. (2015). Mood and personality predict improvement in creativity due to meditation training. *Learning and Individual Differences, 37*, 217–221.

Evans, D. E., & Rothbart, M. K. (2007). Developing a model for adult temperament. *Journal of Research in Personality, 41*(4), 868–888.

Eysenck, H. J., Barrett, P., & Eysenck, S. B. G. (1985). Indices of factor comparison for homologous and non-homologous personality scales in 24 different countries. *Personality and Individual Differences, 6*(4), 503–504.

Friedman, H. S., & Kern, M. L. (2014). Personality, well-being, and health. *Annual Review of Psychology, 65*, 719–742.

Graziano, W. G., & Tobin, R. M. (2009). Agreeableness. In M. R. Leary, & R. H. Hoyle (Eds.), *Handbook of individual differences in social behavior* (pp. 46–61). New York: The Guilford Press.

Grossman, P. (2011). Defining mindfulness by how poorly I think I pay attention during everyday awareness and other intractable problems for psychology's (re) invention of mindfulness: Comment on Brown et al. *Psychological Assessment, 23*, 1034–1040.

Gully, S., & Chen, G. (2010). *Individual differences, attribute-treatment interactions, and training outcomes. Learning, Training, and Development in Organizations* (pp. 3–64). New York: Routledge/Taylor & Francis Group.

Gutiérrez, J. L. G., Jiménez, B. M., Hernández, E. G., & Pcn, C. (2005). Personality and subjective well-being: Big five correlates and demographic variables. *Personality and Individual Differences, 38*(7), 1561–1569.

Hanley, A. W., & Garland, E. L. (2017). The mindful personality: A meta-analysis from a cybernetic perspective. *Mindfulness, 8*(6), 1456–1470.

Jha, A. P., Stanley, E. A., Kiyonaga, A., Wong, L., & Gelfand, L. (2010). Examining the protective effects of mindfulness training on working memory capacity and affective experience. *Emotion, 10*(1), 54.

Keng, S. L., Smoski, M. J., & Robins, C. J. (2011). Effects of mindfulness on psychological health: A review of empirical studies. *Clinical Psychology Review, 31*(6), 1041–1056.

Kiken, L. G., Garland, E. L., Bluth, K., Palsson, O. S., & Gaylord, S. A. (2015). From a state to a trait: Trajectories of state mindfulness in meditation during intervention predict changes in trait mindfulness. *Personality and Individual Differences, 81*, 41–46.

Killingsworth, M. A., & Gilbert, D. T. (2010). A wandering mind is an unhappy mind. *Science, 330*(6006), 932-932.

Komarraju, M., & Karau, S. J. (2005). The relationship between the big five personality traits and academic motivation. *Personality and Individual Differences, 39*(3), 557–567.

Kozlowski, S. W., Gully, S. M., Brown, K. G., Salas, E., Smith, E. M., & Nason, E. R. (2001). Effects of training goals and goal orientation traits on multidimensional training outcomes and performance adaptability. *Organizational Behavior and Human Decision Processes, 85*(1), 1–31.

Lahey, B. B. (2009). Public health significance of neuroticism. *American Psychologist, 64*(4), 241.

Lamers, S. M., Westerhof, G. J., Kovács, V., & Bohlmeijer, E. T. (2012). Differential relationships in the association of the Big Five personality traits with positive mental health and psychopathology. *Journal of Research in Personality, 46*(5), 517–524.

McCrae, R. R., & Costa, P. T., Jr (1985). Openness to experience. *Perspectives in Personality, 1*, 145–172.

McCrae, R. R., & Costa, P. T., Jr (2003). *Personality in adulthood: A five-factor theory perspective*. Guilford Press.

McCrae, R. R., Costa, P. T., Jr, Ostendorf, F., Angleitner, A., Hřebíčková, M., Avia, M. D., … Saunders, P. R. (2000). Nature over nurture: Temperament, personality, and life span development. *Journal of Personality and Social Psychology, 78*(1), 173.

McCrae, R. R., & John, O. P. (1992). An introduction to the five-factor model and its applications. *Journal of Personality, 60*(2), 175–215.

McCrae, R. R., & Costa, P. T. (1987). Validation of the five-factor model of personality across instruments and observers. *Journal of Personality and Social Psychology, 52*(1), 81.

Mussell, M. P., Mitchell, J. E., Crosby, R. D., Fulkerson, J. A., Hoberman, H. M., & Romano, J. L. (2000). Commitment to treatment goals in prediction of group cognitive–behavioral therapy treatment outcome for women with bulimia nervosa. *Journal of Consulting and Clinical Psychology, 68*(3), 432.

Nyklíček, I., & Irrmischer, M. (2017). For whom does mindfulness-based stress reduction work? Moderating effects of personality. *Mindfulness, 8*(4), 1106–1116.

Nyklíček, I., van Beugen, S., & Denollet, J. (2013). Effects of mindfulness-based stress reduction on distressed (type D) personality traits: A randomized controlled trial. *Journal of Behavioral Medicine, 36*, 361–370.

Ortner, C. N., Kilner, S. J., & Zelazo, P. D. (2007). Mindfulness meditation and reduced emotional interference on a cognitive task. *Motivation and Emotion, 31*(4), 271–283.

Park, T., Reilly-Spong, M., & Gross, C. R. (2013). Mindfulness: A systematic review of instruments to measure an emergent patient reported outcome (PRO). *Quality of Life Research, 22*, 2639–2659.

Pervin, L. A., & John, O. P. (Eds.), (1999). *Handbook of personality: Theory and research*. Elsevier.

Rothbart, M. K. (2007). Temperament, development, and personality. *Current Directions in Psychological Science, 16*(4), 207–212.

Ruocco, A. C., & Wonders, E. (2013). Delineating the contributions of sustained attention and working memory to individual differences in mindfulness. *Personality and Individual Differences, 54*(2), 226–230.

Sanderson, C., & Clarkin, J. F. (1994). Use of the NEO-PI personality dimensions in differential treatment planning. In P. T. Costa, Jr., & T. A. Widiger (Eds.), *Personality disorders and the five-factor model of personality* (pp. 219–235). Washington, DC: American Psychological Association.

Schmertz, S. K., Anderson, P. L., & Robins, D. L. (2009). The relation between self-report mindfulness and performance on tasks of sustained attention. *Journal of Psychopathology and Behavioral Assessment, 31*(1), 60–66.

Shahar, B., Britton, W. B., Sbarra, D. A., Figueredo, A. J., & Bootzin, R. R. (2010). Mechanisms of change in mindfulness-based cognitive therapy for depression: Preliminary evidence from a randomized controlled trial. *International Journal of Cognitive Therapy, 3*(4), 402–418.

Shapiro, S. L., Brown, K. W., Thoresen, C., & Plante, T. G. (2011). The moderation of mindfulness-based stress reduction effects by trait mindfulness: Results from a randomized controlled trial. *Journal of Clinical Psychology, 67*(3), 267–277.

Stawarczyk, D., Majerus, S., Van der Linden, M., & D'Argembeau, A. (2012). Using the day-dreaming frequency scale to investigate the relationships between mind-wandering, psychological well-being, and present-moment awareness. *Frontiers in Psychology, 3*, 363.

Studer-Luethi, B., Jaeggi, S. M., Buschkuehl, M., & Perrig, W. J. (2012). Influence of neuroticism and conscientiousness on working memory training outcome. *Personality and Individual Differences, 53*(1), 44–49.

Tang, Y. Y., Hölzel, B. K., & Posner, M. I. (2015). The neuroscience of mindfulness meditation. *Nature Reviews Neuroscience, 16*(4), 213–225.

Talbot, N. L., Duberstein, P. R., Butzel, J. S., Cox, C., & Giles, D. E. (2003). Personality traits and symptom reduction in a group treatment for women with histories of childhood sexual abuse. *Comprehensive Psychiatry, 44*(6), 448–453.

Thomson, P., Jones, J., Browne, M., & Leslie, S. J. (2014). Psychosocial factors that predict why people use complementary and alternative medicine and continue with its use: A population based study. *Complementary Therapies in Clinical Practice, 20*(4), 302–310.

Wheeler, M. S., Arnkoff, D. B., & Glass, C. R. (2016). What is being studied as mindfulness meditation? *Nature Reviews Neuroscience, 17*(1), 59.

Zuroff, D. C., Koestner, R., Moskowitz, D. S., McBride, C., Marshall, M., & Bagby, M. R. (2007). Autonomous motivation for therapy: A new common factor in brief treatments for depression. *Psychotherapy Research, 17*(2), 137–147.

Chapter 3

Cultural differences in meditation

Culture involves a set of collective social behavior and norms, beliefs, attitudes, values, and traditions. A new research field, cultural neuroscience, incorporates neuroscience into cultural research and has examined how cultural and genetic diversity shape the human mind/brain, body (physiology), and behavior using multiple methods involving behavior, neuroimaging, physiology, and genetics (Chiao, Cheon, Pornpattananangkul, Mrazek, & Blizinsky, 2013). Using these multimodal and multilevel methods, a large body of research have shown cultural differences in perception, attention, cognition, emotion, and social behavior in healthy and patient populations.

Cultural differences in attention and perception

Most fundamental psychological processes such as perception were previously assumed to be universal, but growing evidence suggests that they can be affected by culture. Using diverse tasks, studies indicated that when American and Chinese participants were exposed to the same environment or context, Americans focused more on salient and focal objects whereas Chinese focused more on background (Nisbett & Miyamoto, 2005). Further studies that tracked eye movements also detected the same attention and perception patterns. For example, participants rated how much they liked each picture when they were presented to a series of pictures including a focal object (e.g., a tiger) on a background (e.g., the jungle). Eye tracking results showed that Americans looked at the focal object sooner and fixated on it longer than Chinese. In contrast, Chinese participants made more rapid eye movements from one location to the next, including the object and background. These studies suggested cultural differences in attention and perception, such that Americans pay more attention to objects and their detailed features, consistent with an analytic processing style, whereas Chinese attend more to contextual relationship, reflecting holistic processing (Chua, Boland, & Nisbett, 2005; Goh & Park, 2009).

Based on this growing evidence, a review article concluded that perceptual and attentional processes are influenced by cultures. Americans attend to

The Neuroscience of Meditation. DOI: https://doi.org/10.1016/B978-0-12-818266-6.00004-6

context-independent and analytic perceptual processes via focusing on a salient object independently, whereas Chinese attend to context-dependent and holistic perceptual processes through the object—context relationship. The underlying mechanisms of cultural differences may derive from daily experiences and participation in different social practices that lead to the adaptation and changes in perception and attention. In other words, different cultures produce different characteristics and default patterns for perception and attention (Nisbett & Miyamoto, 2005). These findings demonstrated a dynamic relationship between the cultural context and perceptual processes and suggested that perception is not a universal process that is identical across all cultures. People from Western and Eastern cultures attend, perceive, and receive information significantly differently even when they are placed in the same environment and context, suggesting fundamental differences in perception, attention, and subsequent cognitive processes between the two cultures (Nisbett, 2004; Tang, 2017).

Do cultures affect brain processes related to attention and attention control? Behavioral research has indicated that people from Western cultures (e.g., Americans) perform better on tasks emphasizing independent (absolute) dimensions than tasks focusing on interdependent (relative) dimensions, whereas people from Eastern cultures (e.g., Chinese) behave reversely. A study using Chinese and American participants used functional magnetic resonance imaging to examine how their brains responded during simple visuospatial tasks——absolute judgments (ignoring context) or relative judgments (considering context). Results showed that brain activations in frontal and parietal areas associated with attention control were greater during culturally nonpreferred judgments than during culturally preferred judgments. In other words, the Chinese participants showed more brain activations in absolute judgments (ignoring context) whereas American participants had more brain activations in relative judgments (considering context). Moreover, brain activation differences correlated with individual differences in culture-typical identity in each group. These findings suggest that even for simple visuospatial tasks, cultural differences moderate brain activations related to attention and attention control (Hedden, Ketay, Aron, Markus, & Gabrieli, 2008). Together these findings also shed light into the learning and practice of meditation given that meditation is a systematic training of attention and self-control; an intervention that is deeply rooted in culture and tradition (Tang, 2017; Tang, Holzel, & Posner, 2015).

Attention patterns and control are some of the most important processes related to meditation practice and outcomes. For instance, if Americans pay more attention to salient and focal objects and their detail, these analytic perceptual processes may lead to preference for a focus or concentration strategy during meditation. However, it is difficult to conclude whether Americans are good at focusing or concentration-based meditation and would benefit from it easily without empirical support. Conversely, it is also likely that a too narrow focus may result in a stressful state of attention control, which could interfere with meditation practice and outcomes. Based on teaching integrative body−mind training (IBMT)——a form of mindfulness meditation in children and adults—our experiences suggest that a narrow focus often induces more stressed and survival modes of attention control because participants often put too much effort into manipulating attention and controlling their minds, which could lead to imbalanced body−mind states. However, if participants use an open focus strategy to attend to an object, they could perform concentration-based meditation well and benefit from it (Tang, 2017). If people from independent cultures such as the United States are trained to see more of the context and relations between objects and context, what would happen to their meditation practice and outcomes? We conducted several randomized controlled trials (RCTs) to examine IBMT mechanisms and applications in college students in the United States and our observations suggested that Americans benefit from this context-based strategy because they could have two types of attention patterns (narrow vs open focus), which help increase mental flexibility during meditation practice. However, we still need more data to determine to what extent the strategy may affect meditation outcomes.

Cultural differences in somatic or interoceptive awareness

Cultures shape attention and perception and may affect meditation practice and outcomes. Given that attention to, and perception of, bodily sensation and states are important to meditation practice, a key question related to cultural differences concerns our bodily sensations and body awareness. To what extent do people from different cultures vary in their somatic or interoceptive awareness?

Somatic or interoceptive awareness often refers to identifying internal physiological processes related to bodily sensations, cognitive processes, and emotions (Craig, 2015). Studies have shown cultural differences in somatic or interoceptive awareness. For example, a review article explored cross-cultural differences in somatic awareness and interoception and the role of culturally bound epistemologies and contemplative practices (Ma-Kellams, 2014). Cross-cultural differences in interoception suggest that: (1) people from Eastern cultures often show higher levels of somatic awareness, but lower levels of interoceptive accuracy; (2) differences in cultural conceptualizations and epistemic traditions partially explain these differences; (3) meditation facilitates bodily awareness; and (4) the heightened somatic awareness among Eastern cultures is linked to greater emphasis on somatic symptoms in diverse psychopathologies——most notably anxiety and depression.

Since somatic or interoceptive awareness is directly related to meditation practice and outcomes, it is important to improve somatic or interoceptive awareness through training. In general, somatic-awareness training aims to improve awareness of the body. It often combines deep or abdominal breathing and body movement to gradually increase internal body sensations (e.g., feeling tension in the muscle groups). In fact, there are better ways to train and improve somatic awareness. For example, in our IBMT translational work we use bodifulness and mindfulness to improve somatic or interoceptive awareness (Tang & Tang, 2015a, 2015b; Tang, Tang, & Gross, 2019). In IBMT practice, cooperation between body and mind is emphasized in facilitating and achieving a meditation state as well as better training outcomes (Tang, 2009, 2017; Tang et al., 2007, 2015). Our series of RCTs have shown that a key mechanism of IBMT involves the interaction between the central nervous system (CNS, brain) and autonomic nervous system (ANS, body; Tang et al., 2007, 2015, 2019; Tang, 2009). In addition to mindfulness practice, IBMT also uses bodifulness to achieve better outcomes. Bodifulness refers to the gentle adjustment and exercise of body posture with full awareness in order to achieve presence, balance, and integration in our bodies. Bodifulness not only exercises muscle groups, but also engages internal organs, which are important components leading to visceral and interoceptive processes. Therefore bodifulness mainly involves implicit processes such as visceral and interoceptive awareness regulated by ANS. Autonomic control exerted by ANS requires less effort and is mainly supported by the anterior cingulate cortex/medial prefrontal cortex (ACC/mPFC) and striatum (Critchley et al., 2003; Tang, 2017; Tang et al., 2015, 2019).

Full awareness and presence of the body (bodifulness) could facilitate the mindfulness process, consistent with the literature that body posture and state affect mental processes such as emotional processing, retrieval of autobiographical memories, and cortisol concentrations (Tang et al., 2019). We developed a novel bodifulness technique for IBMT that can directly train the presence, balance, and interaction within internal organs and visceral processing rather than just body muscle groups. The major difference between IBMT and general somatic training is whether the training could trigger and induce a physiological response and sensation that could guide participants to naturally attend to and engage in these inner sensations that are similar to acupuncture sensations—Asian cultures often call it "Qi"—along energy channels or meridians (invisible lines of energy flow in traditional Chinese medicine). Are these sensations just imagination? We conducted studies using a German Vega Whole Body System (an electrodermal measurement equipment) to measure whole body changes indexed by the skin's electrical resistance and electrodermal screening. It seemed that these feelings were different from imagination, but further rigorous research is warranted (Tang, 2009).

One interesting phenomenon we found during IBMT learning and practice is that better somatic or interoceptive awareness can signal and stimulate our bodies to regulate and rebalance our physiological system naturally, indicating that self-rehabilitation or self-healing mechanisms are involved during these bodily regulatory processes. For instance, after practice, IBMT practitioners reported: "I often feel an internal but natural urge from within (inside or internal somatic or interoceptive awareness) to change my postures after sitting for hours or working in a fixed posture for longer time. If I still keep the same postures after internal reminder, I would feel uncomfortable. At this moment I have just realized my postures are actually not relaxing and in good state that need a change or adjustment." These reports suggest that behavior change is initiated from internal and implicit processes and move toward external and explicit awareness and processes.

A well-trained body can remind us whether and how to change our body postures and states within. Our series of RCTs in preventing and changing addiction behavior provide scientific evidence supporting these spontaneous behavior changes from within (Tang, 2017; Tang, Tang, & Posner, 2013; Tang et al., 2015).

Based on growing evidence, we developed a model of integrative body–mind training (IBMT) and psychological well-being (Tang et al., 2015, 2019) and concluded that IBMT improves psychological well-being through mindfulness and bodifulness that mainly strengthen self-control abilities and related CNS (i.e., ACC/mPFC and striatum) and ANS systems (i.e., parasympathetic activity). Bodifulness significantly facilitates and improves somatic or interoceptive awareness, which helps achieve the meditative state (Tang et al., 2009, 2015, 2019; Tang, Lu, Geng, Stein, Yang, Posner, 2010; Tang, Lu, Fan, Yang, & Posner, 2012).

Cultural differences in reasoning and decision-making

Reasoning and decision making, as important aspects of a person's high-level cognition, are also influenced by cultural differences. For example, Asians tend to frame the decision to help as a matter of moral responsibility, whereas Americans are likely to take it more as one's personal choice (Wang, Deng, Sui, & Tang, 2014). In this section we discuss the cultural differences between Chinese and American cultures in abstract thinking/reasoning and moral decision making, which are related to the cultivation of a meditative state.

Eastern holistic systems of thought rely on connectedness and relations as a primary way of understanding the world, whereas Western analytic systems rely on discreteness as an epistemological way of thinking (Nisbett, Peng, Choi, & Norenzayan, 2001). For example, a popular national TV competition called Sing! China started in 2016 and each season has attracted audiences of over 10 million. Each singer is judged by four experts and the singer could select one expert as his or her singing coach. When more than one expert (e.g., 2, 3, or 4) favors the singer, the singer must choose one of them. This process involves thinking, reasoning, and decision making.

Interestingly, the Chinese singers often didn't make decisions solely based on their opinion or choice, but rather considered interdependence, connectedness, and relationships.

In fact, many singers made decisions based on their parents' favorite expert or by peer influence, suggesting the high impact of collectivism culture. In contrast, in similar TV competition called The Voice of Holland, singers from the individualism culture mainly made decisions based on their own opinions. Relatedly, decisions concerning meditation practice in Chinese culture could be affected by family members such that children are influenced by parents and adults are affected by partners or friends. These observations are consistent with research findings in cultural differences. We discuss arithmetic processing as an example of abstract thinking/reasoning as well as moral decision making to demonstrate the neural basis of cultural influences on these processes that involve Eastern and Western cultures (Tang & Liu, 2009).

Arithmetic often involves abstract thinking and reasoning. One example of a higher cognitive function that may be influenced by holistic versus analytic thinking styles involves discrete neural patterns of arithmetic processing in native Chinese and English speakers (Tang et al., 2006). The universal use of Arabic numbers in mathematics raises the question of whether they are processed the same way by people from different cultures and who speak different languages. To address this question, we used functional magnetic resonance imaging (fMRI) to scan native Chinese speakers (NCS) and native English speakers (NES) who have college-level education. The participants were instructed to perform four tasks during the brain scan. (1) Symbol condition—judging the spatial orientation of nonnumerical stimuli in which a triplet of nonsemantic characters or symbols are visually presented either in an upright or italic orientation. This simple perceptual task was to decide whether the third symbol had the same orientation as the first two. (2) Number condition—judging the spatial orientation of numerical stimuli using a similar perceptual task as in the symbol condition, but using Arabic digits as visual stimuli. (3) Addition condition—a numerical addition task was to determine whether the third digit was equal to the sum of the first two in a triplet of Arabic numbers. (4) Comparison condition—a quantity comparison task to determine whether the third digit was larger than the first two in a triplet of Arabic numbers. We controlled the digits (1−9) or the sum of digit addition to be under 9 in conditions 2, 3, and 4, thus the tasks were easy for participants. A baseline condition of matching white and/or gray circular dots was used to control for the motor and nonspecific visual components of the tasks (Tang et al., 2006).

Behaviorally there were no significant differences between NES and NCS in all four conditions, suggesting the same difficulty levels of the tasks. We found slight brain differences in condition (1) with nonsemantic characters or symbols between NES and NCS, but the difference were not significant. Given that nonsemantic characters or symbols are simple and only involved basic and similar perceptual processing, these results showed similar brain activations. If we increase the complexity of the characters or symbols, we may be able to detect brain differences in these mental processes cross-culturally. However, in condition (2), the perceptual digit task without any mental calculation, our results showed very different brain activations in the cortical representation of numbers between NES and NCS. These results suggest distinctive engagements in the brain between Western and Eastern cultures even in a simple perceptual task involving digits, consistent with behavioral research in cultural differences in perception and attention. In condition (3), the simple addition task, the English speakers employed a language process relying on the Broca and Wernicke language areas for mental calculation, whereas the Chinese speakers engaged a visuo-premotor association network (between BA6, BA8, and BA9) for the same task, suggesting culture affect brain processing even in simple addition. We further chose two regions of interest——the Broca and Wernicke language areas and the premotor association area—to conduct quantitative analyses by comparing the fMRI signal changes between the English and Chinese groups. We found that Broca and Wernicke activations were significantly greater in the English speakers than in the Chinese speakers, suggesting NES use language brain areas to help math calculation.

As the arithmetic complexity increased across the four conditions (symbol < number < addition < comparison), there was a trend toward increased premotor activation in the Chinese speakers, but not in the English speakers. Therefore there was a double dissociation in brain activation between these two groups during these tasks, supporting clear cultural differences in the processing of numbers. In addition, for both cultural groups the inferior parietal cortex was activated in the numerical quantity comparison; however, the fMRI connectivity analyses revealed a difference in the brain networks during the task for Chinese and English speakers. Two distinct patterns were observed in the functional networks: there was dorsal visuo-pathway dominance (through the parietal−occipital cortex) for the Chinese speakers, but ventral visuo-pathway dominance (through the temporal cortex) for the English speakers, suggesting that cultures shape the brain and affect behavior. Our findings have two implications. First, in both Chinese and English speakers, there is a cortical dissociation between addition and numerical comparison processing. The addition task seems to be more dependent on language processing than does the comparison task, which is consistent with the notion that there are different neural substrates underlying verbal and numerical processing. Second, there are differences in the brain representation of number processing between Chinese and English speakers (Tang & Liu, 2009).

It should be noted that Chinese is a logographic language, where a single character is represented phonetically and ideographically. In contrast,

English is an alphabetical language and uses letters to represent sounds only, rather than entire concepts. Previous research has shown brain differences in the processing of these two languages, and language could be one of the critical cultural factors affecting numerical cognition as each culture typically has a distinct way of pronouncing and representing numbers. For Chinese speakers, the acquisition of language depends heavily on learning logographic characters and memorizing each stroke and subcharacter comprising individual characters, which could result in repeated activation of the visuomotor association brain area. On the other hand, English is relatively phonetic, which may explain why Broca's area is more engaged since Broca's area is associated with speech production (Tang et al., 2006, 2009). Chinese reading has been suggested to rely on areas outside the classical network for alphabetic reading such as Broca and Wernicke areas. Historically, Chinese children learn to read through handwriting which associates the visuographic properties of characters with lexical meaning. Nowadays Chinese children learn to use electronic devices based on the pinyin input method, which links phonemes and English letters to characters (e.g., closer to English language learning). A study showed that this significant change of learning processes affected children's reading scores such that reading ability was negatively correlated with their use of the pinyin input method, suggesting that in the digital age, pinyin typing on e-devices hindered Chinese reading development (Tan, Xu, Chang, & Siok, 2013). Related to our results, the different biological encoding of numbers may be shaped by visual reading experience during language acquisition and other cultural factors such as mathematics learning strategies and education systems, which cannot be explained completely by differences in languages per se (Tang et al., 2006).

Although differences between the two cultures in terms of language, educational systems, or genetics may contribute to these findings, there is a relatively greater reliance on analytic and logical (rather than holistic and contextual) styles of cognition among Westerners, which may contribute to the greater observed activity in language-related regions during the arithmetic and other cognitive tasks such as reasoning and decision making. As a result these analytic and logical styles of mental processes could influence meditation decisions and practice. Future cultural neuroscience research should investigate the consequences of holistic versus analytic forms on thought as well as on diverse cognitive processes, which would help to confirm the culturally variant patterns of neural activity observed during these tasks. Furthermore, genetics and environmental factors often work together in shaping human cognition, emotion, and social behavior (Tang, 2017; Tang & Liu, 2009). For example, if both genes and experience shape human cognitive functions such as numerical processing in the brain, it is important to understand how different educational systems may influence our core cognitive capacities (Tang & Liu, 2009). Future studies should address how gene × environment (or experience) interact to affect high-level cognition such as decision-making in different cultures and how to optimize reasoning and decision-making abilities through learning from other cultures.

Moral reasoning and decision making have attracted much scholarly attention. The growing literature has tried to uncover the brain mechanisms underlying moral decision making with a family of moral dilemmas (Greene, Nystrom, Engell, Darley, & Cohen, 2004). Generally a moral dilemma is a complex situation that involves conflict in choosing between two undesirable alternatives, which could evoke the competition between deontological (non-utilitarian) choice and utilitarian response (Wang et al., 2014). For example, in a personal moral dilemma (the footbridge dilemma), the only way to save five workers from a runaway trolley is to push a large man off an overpass bridge onto the tracks below. He will die, but his body will stop the trolley from reaching the other five people. A corresponding impersonal moral dilemma is the trolley dilemma, in which the only way to save the five workers is to pull a lever redirecting the trolley onto another set of tracks, where it will kill a single worker instead of five workers (Greene et al., 2004; Wang et al., 2014). The deontological response is an aversive emotional response to the harmful act, which would lead to the rejection of utilitarian response. In contrast, the utilitarian response is to take part in the harmful act since doing so will maximize good consequences, which would require overcoming the prepotent emotional response.

Although the proposed actions in the personal and impersonal dilemmas would produce similar outcomes, moral judgment in the two dilemma types might be driven by different principles. Previous studies have indicated that most people show agreement with pulling the lever in the trolley dilemma and disagreement with pushing the man in the footbridge dilemma (Greene et al., 2004; Wang et al., 2014). Neuroimaging results showed that personal moral dilemmas elicit greater activation in brain regions associated with emotions, whereas impersonal moral dilemmas elicit greater activation in areas associated with problem solving and working memory, suggesting that cognitive and emotional processes contribute to moral decision making (Greene et al., 2004). For personal and impersonal ethical dilemmas, each act may lead to certain consequences and either side may be right in different circumstances. If we adopt the utilitarian way of thinking, we would conclude that it is right to kill one person instead of five people, but it is also right to develop an intuitive rule against the participation in killing others. However, people's moral decisions may be influenced by cultural factors, since people's morals and virtuousness are shaped by culture (Rozin, 2003).

Preliminary research on moral decision making suggests that people from holistic cultures (Chinese) exhibit greater integrative processing of moral dilemmas than those from analytic cultures (Americans). One of our studies (culture × type of dilemma) combined event-related potential (ERP) techniques with standardized low-resolution brain electromagnetic tomography (sLORETA) to study potential cultural variation (Wang et al., 2014).

Chinese and American college students were recruited to participate in a moral-dilemma task that included both personal and impersonal situations. An example of a personal situation is the footbridge dilemma, while an impersonal situation includes the trolley dilemma. Each dilemma was presented in the participants' native language (Chinese or English) and as black text against a gray background on a computer monitor, with a series of three screens. The first two screens described the scenario of the dilemma and the third posed a question asking whether or not the hypothetical action was morally appropriate. Choosing appropriate options was considered to be utilitarian (taking part in the harmful act since doing so will maximize good consequences, i.e., saving five workers in the footbridge dilemma, which requires overcoming the prepotent emotional response), whereas choosing inappropriate options was considered to be nonutilitarian. The dependent variables were the proportion of utilitarian judgments, reaction time, ERP responses, and sLORETA brain activity.

We detected a main effect of type of dilemma such that participants made a smaller proportion of utilitarian judgments and exhibited longer reaction times in response to personal than impersonal dilemmas, with no difference observed between Americans and Chinese. We focused the P3 and P260 ERP components, brain indexes of the process of decision-making. Notably, the ERP components were significantly different between the two cultural groups. For Americans, smaller P3 amplitudes were evoked by personal rather than impersonal dilemmas, whereas for Chinese, smaller P260 deflections were elicited by personal rather than impersonal dilemmas. The different ERP components elicited by moral dilemmas may be attributable to cultural differences as it has been widely demonstrated that these two cultures differ in experience and socialization, which also influences cognition and the allocation of attention (Kitayama & Park, 2010). Previous research has indicated that P3 is an index of inhibition of task-irrelevant emotional information, with less positive amplitudes for negative stimuli than neutral stimuli in implicit emotional tasks. Our P3 results are consistent with prior findings, suggesting that more negative emotions are needed to be inhibited in response to personal rather than impersonal dilemmas. Moreover, the P260 component has been reported to reflect immediate affective reactions toward options that integrate attention, working memory, and emotional processing, thus the different cognitive functioning processes may be related to holistic thinking, suggesting a more integrated process during the solution of moral dilemmas in Chinese compared to Americans.

The sLORETA analysis showed a significantly different pattern of brain activity between Americans and Chinese. The main source of both P2 and P3 components for personal dilemmas was the cingulate gyrus, similar to other studies (Gajewski, Stoerig, & Falkenstein, 2008). In contrast, the main sources of P2 and P3 components for impersonal dilemmas were localized in the medial frontal area and cingulate gyrus (with contributions from several other brain regions, including the temporal and insula areas). These findings are in line with brain imaging research demonstrating that a complex network of brain regions is involved in moral decision making (Prehn et al., 2008). Different from the sources of the P2 and P3 components, the P260 component for both dilemma types was mainly activated in areas in the posterior cingulate, parahippocampal gyrus, and cuneus and precuneus cortices, and these areas are considered to be related to emotional processing and evaluation, retrieval of episodic memory representations, and attention, as well as the detection of salient stimulus and higher-order cognitive functions (Cavanna & Trimble, 2006; Daniel & Donna Rose, 2007). These results suggested that in the Chinese participants, brain areas associated with attention, memory retrieval, and emotional processing were involved in the process of moral decision making and may also involve holistic thinking. Future research examining cultural differences in these processes would benefit from direct manipulations of holistic and analytic thinking.

In summary, we discuss abstract thinking/reasoning and moral decision making in Eastern (e.g., Chinese) and Western (American) cultures to reveal the underlying brain mechanisms using the ERP technique and fMRI. This evidence suggests extensive differences in cognitive and affective processing and moral decision behavior between Eastern and Western cultures, and that differences in the resolution of moral dilemmas would be reflected in spatio-temporal cortical activation underlying moral decision making. These results using diverse methods support a dual-process theory of moral judgment, which states that utilitarian moral judgments (favoring the "greater good" over individual rights) are enabled by controlled cognitive processes, while deontological judgments (favoring individual rights) are driven by intuitive emotional responses. Future studies will need to address how cultural differences in these moral reasoning and decision-making processes may influence meditation practice and outcomes.

Cultural differences in social behavior and self-related process

One influential explanation for our findings is that besides potential differences in biology (e.g., genes) across cultures, social factors also play an important role in directing attention, perception, decision making, and behavior. For example, Asians often live in a complex social system that involves large families with relatives living together, thus attention to context is important to effective functioning in daily life. Moreover, the external environments of the East are also far more complex and compacted, containing more people and objects such as high rises and public infrastructure than those of the West, which further influences perception, cognition, emotion, and social behavior (Chiao, Li, Seligman, & Turner, 2016; Nisbett, 2004).

Social orientation such as independent versus interdependent is one important social behavior. Previous research suggests that Americans are more independent (less interdependent) than Asians, making these cultural groups vary in terms of social orientation. In other words, Americans place a strong normative emphasis on independence of the self from others, whereas Asians put more emphasis on interdependence of the self with others. These cultural differences show behavioral cultural traits such as more holistic attention and perception, more emphasis on social (vs personal) happiness, and greater suspension of self-interest, which are all associated with more interdependent orientations (Kitayama et al., 2014; Kitayama & Park, 2010; Nisbett, 2004; Nisbett & Miyamoto, 2005). These different social orientations of independence versus interdependence are thought to result from the acquisition of cultural norms and values. Most importantly, the fundamental differences in perception and attention (e.g., holistic and contextual vs analytic and logic styles) could affect receiving and processing information momently and significantly contribute to social behavior. Another important topic is individualism and collectivism (e.g., Chinese vs American cultures), which refer to cultural values that influence how people construe themselves (e.g., self-related process) and their relation to the world. People from individualist cultures perceive themselves as stable entities, autonomous from others and environment, whereas people from collectivist cultures view themselves as dynamic entities, continually influenced by their social context and relationships, which is consistent with the cross-cultural evidence in perception, attention, thinking, and decision making. Of note, studies in Chinese and American samples also suggested that individualism and collectivism have influences on individual differences in self-control (Li, Vazsonyi, & Dou, 2018). Given that self-control is crucial in meditation practice, these findings shed light onto how cultures influence self-control, a key capacity that meditation practice aims to cultivate (Tang, 2017).

"Self" is a set of thoughts, beliefs, feelings, memories, values, and behavioral predispositions that we have bought into and integrated into a relatively stable verbal picture of ourselves. Culture plays a crucial role in ascribing meaning and value to the self and the self-related process. People in different cultures have different construals of the self and others, which influence individual experiences in cognition, emotion, attitude, and motivation. For instance, Chinese culture has distinct conceptions of individuality and emphasizes the relatedness of individuals to each other such as attending to others, fitting in, and having harmonious interdependence with them. American cultures neither assume nor value the connectedness among individuals, instead, individuals seek to maintain their independence from others by focusing on, and expressing, themselves. Many cultural differences are based on the ways that people across cultures construe and represent the self, others, and the world (Kim & Sasaki, 2014). In summary, culture shapes the self-related processes such as cognition, emotion, motivation, belief, and behavior. Cultural differences in understanding and conceptualizing one's relationships with others have led to diverse cultural systems for interpreting, thinking, feeling, reasoning, and decision making about the world and others (Kim & Sasaki, 2014; Kitayama & Park, 2010; Markus & Kitayama, 1991).

Meditation practice involves at least three mental processes such as attention control, emotion regulation, and altered self-awareness. Through practice, a meditator could experience a systematic change in body, mind, behavior, and, most importantly, self-awareness (Tang et al., 2015). Self-awareness mainly involves monitoring and being aware of our inner thoughts, feelings, emotions, memories, values, and beliefs, which are actually the "self." The practice evokes a metacognitive state that transforms how we attend to our experiences. Beyond the self as a process of ongoing self-awareness, the meditator often experiences reconstructing and deconstructing the self and then moves toward observing the self, transcendent self, or other concepts of the true self. Although we have not fully understood these unique human experiences of the self, it seems that a higher level of self-awareness, equanimity, self-transcendence, and self-transformation has been developed and the understanding of the self, self−others, and self−nature has dramatically changed, suggesting meditation practice does change the self and self-related processes (Tang & Tang, 2014, 2015a, 2015b). Given that self-related processes are also shaped by Eastern and Western cultures differently, it will be interesting and important to learn how cultures (in the East and West) and meditation practice interact to change self and self-related processes and, thus, influence human development. For example, Americans tend to emphasize meditation techniques to achieve specific states such as Jhana, whereas Chinese pay more attention to spontaneous wisdom induced by practice, consistent with the findings of the cultural differences in analytic versus holistic perception, attention, and thinking. These cultural differences may further affect attitude, motivation, belief, and mindset of meditation practice, strategies, and outcomes. Therefore open-minded learning from each culture becomes more crucial for meditation practice and benefits (Tang, 2017).

Take social cognition as an example. Despite the rich understanding of how individualism and collectivism influence social cognition in behavior, little is known about how these cultural values modulate neural activity underlying social cognition. A cross-cultural fMRI study examined how individualism and collectivism modulate neural activity in mPFC during processing of general and contextual self-judgments. Results showed that the anterior rostral part of the mPFC activity during self-judgments positively predicted how individualistic or collectivistic a person is across cultures. These results suggested two kinds of neural representations of self, a general self and a contextual self within mPFC, and demonstrated how cultural values of individualism and collectivism shape these neural representations (Chiao et al., 2009). A quantitative meta-analysis of fMRI studies on cultural differences indicated that Eastern cultures are associated with increased brain activity in the regions related to inference of others' mind and emotion regulation, whereas Western cultures are related to enhanced brain activity in areas related to self-relevance encoding and emotional responses during social cognitive and affective processes (Han & Ma, 2014). Meditation research has provided further evidence. For instance, self-referential processes often activate cortical midline structures (CMS), but mindful self-awareness is related to decreased CMS activation. An fMRI study investigated experienced mindfulness meditators and matched meditation-naïve participants during mindful self-awareness (FEEL) and self-referential thinking (THINK) conditions. As expected, decreased activations in CMS were detected during FEEL for both groups, but meditators showed significantly stronger decreases in prefrontal CMS (Lutz, Brühl, Scheerer, Jäncke, & Herwig, 2016).

We have discussed cultural differences based on brain functional changes such as activity and connectivity. Do different cultural contexts affect different brain structures? Some studies on NES and Chinese speakers have suggested that there are neuroanatomical differences between cultural groups. Culture and language are common explanations for the structural differences, but genetic and environmental factors should also be considered. It should also be noted that cross-cultural differences are not static, but dynamically consistent due to the chronic and moment-to-moment salience of individualism and collectivism (Oyserman & Lee, 2008). The results in self-related processes and social behavior are in line with gene-culture coevolution theory such that cultures and genes always work together to drive our representations of self, others, and social behavior (Kim & Sasaki, 2014).

These findings provide insights into meditation learning and practice. For example, people in collectivistic societies (such as Chinese) may be more open to external experience and environment and willing to follow group-based thinking, decision, and action, which is consistent with the cultural values of collectivism and interdependence. Our meditation studies suggested that older Chinese adults enjoyed group practice, interaction, and support more than younger adults. In contrast, Americans behave differently and prefer more individual practices (Tang, 2009). When Americans live in the retirement community with a shared schedule, would it be different from the independent living family? We visited some large retirement communities in the United States where the majority of residents were white Caucasians. Although most residents made friends in the community and followed the same daily schedule, it's different from Chinese behavior and preference, suggesting the effects of individualism and independence. However, it might be possible for Americans in the retirement community to adapt to new norms and change social behavior gradually based on social learning and group influences, but cultural differences would certainly play an important role. What about young adults in China and the United States? Our observations and studies indicated that Chinese and US young adults in meditation behavior appeared to be similar in that either groups or individual sessions work for them. One possible reason might be that the younger generations of Americans and Chinese were born in the digital and globalization age and received less influence from their own traditions and cultures. However, it seems that Chinese and Americans showed differences in following the principles and instructions of meditation practice (Tang, 2017; Tang et al., 2006). Future research should address the gene−culture interaction on meditation behavior and figure out how to learn from different cultures to facilitate meditation learning, practice, and outcomes.

Cultural differences in emotion processes and mental health

One emotion regulation strategy is the suppression of feelings. Studies showed that suppression is more common in Asian cultures than US culture (Kim & Sasaki, 2014; Matsumoto, Yoo, & Nakagawa, 2008). In US culture, expression of emotion is encouraged more strongly. For instance, Americans report experiencing more positive emotions than negative emotions, but this pattern is atypical for Asians. Thus cultures moderate the processes of emotion regulation and the way people feel in response to environment and experience (Kim & Sasaki, 2014). If cultures shape emotion regulation, we should expect these cultural differences also influence related neural activity. One study examined brain mechanisms of disgust suppression in Americans. Although participants reported experiencing reduced negative effects, greater emotion-related responses in the insula and amygdala were detected, suggesting self-report and brain response were discordant (Goldin, McRae, Ramel, & Gross, 2008). A similar study on Japanese did not find increased activation in the insula or amygdala during emotion suppression (Ohira, Nomura, Ichikawa, Isowa, & Iidaka, 2006). Although these studies did not directly compare different cultural groups, the divergent findings within each culture are meaningful. Of note is that cultures not only affect emotion processes, but also work together with genes to moderate emotional experience, well-being, and health (Tompson et al., 2018), see Chapter 4, Genetic association with meditation learning and practice (outcomes) for details.

Emotion processes and regulation are directly related to mental health and disorders, so we expect there are cultural differences in the experience of symptoms and mental disorders. Our experience is associated with attention and interoception that are shaped by cultures. Attentional bias refers to the tendency of our perception to be affected by our recurring thoughts and may explain our failure to consider alternatives, which is also influenced by culture. For example, smokers often have attentional bias for smoking-related cues. Sleep-related attentional bias plays an important role in the development and maintenance of insomnia (Bar-Haim, Lamy, Pergamin, Bakermans-Kranenburg, & van IJzendoorn, 2007; Harris et al., 2015).

Attentional bias is also involved in the exacerbation and maintenance of chronic pain (Rusu, Gajsar, Schlüter, & Bremer, 2019) and is associated with symptoms of disorders such as anxiety, depression, and posttraumatic stress disorder (PTSD) (Browning, Holmes, & Harmer, 2010). In a cross-cultural study of attentional bias in chronic fatigue syndrome, compared to controls, Dutch and British chronic fatigue patients showed a significant attentional bias for illness-related words and were significantly more likely to interpret ambiguous information in a somatic way (Hughes, Hirsch, Nikolaus, Chalder, Knoop, & Moss-Morris, 2018).

A review article examined cross-cultural differences in interoception and the role of culturally bound epistemologies and contemplative practices. The review summarized that people from Western and Eastern cultures exhibit differential levels of interoceptive accuracy and somatic awareness, and show different culturally bound psychopathologies, including somatization, body dysmorphia, pain sensitivity, and eating and mood disorders (Di Lernia, Serino, & Riva, 2016; Ma-Kellams, 2014). One explanation is that culture influences the attention allocation and subsequent cognitive, affective, and social processes. Therefore people in different cultures pay attention differently to the same thing in the same context, which may lead to different mental processes and interpretation of mental health and experience of symptoms and mental disorders (Boduroglu, Shah, & Nisbett, 2009). In summary, these results suggest that culture as a set of social behavior and norms can shape experiences of symptoms and disorders, including attitudes, mindsets, beliefs, and expression of symptoms and diseases, as well as treatment-seeking (Chase, Sapkota, Crafa, & Kirmayer, 2018).

Given that our social cognitive and affective processes are associated with how we believe and perceive mental health, and how we feel, interpret, and experience symptoms and cope with disorders, these research findings provide evidence that culture shapes our belief and attitude toward mental health, influence our feelings and experience of symptoms and mental disorders, and treatment preferences (Jimenez, Bartels, Cardenas, & Alegría, 2013; Jimenez, Bartels, Cardenas, Dhaliwal, & Alegría, 2012; Kramer, Kwong, Lee, & Chung, 2002). For instance, Asian patients tend to report somatic symptoms first and then later describe emotional afflictions, indicating that people selectively present symptoms in a "culturally appropriate" way that won't reflect badly on them (US Department of Health and Human Services, 2001). In Western culture, we are socially asked how we feel and name our emotional states from an early age. Yet this is not the case everywhere such as in some Eastern and African cultures. In reality, whether we describe emotional or physical symptoms depends on our cultural beliefs and norms. Without considering these cultural differences, physicians may misunderstand and misinterpret the symptoms and make an incorrect diagnosis. Patients are often not aware of the cultural influences and only express "culturally appropriate" symptoms, which may mislead the diagnosis

(Chase et al., 2018; Jimenez et al., 2013; Kohrt & Harper, 2008; Kramer et al., 2002). Moreover, some Asian patients prefer avoidance of upsetting thoughts with regard to personal problems rather than explicitly expressing their distress. Compared to white Americans, African Americans are more likely to handle personal problems and distress on their own or turn to their spirituality for support. Moreover, cultural factors often determine how much support we have from our families and communities in seeking help, where people from an Eastern collectivism culture versus Western individualism culture behave very differently (Hunt et al., 2013; Jimenez et al., 2013; U.S. Department of Health and Human Services, 2001). Therefore some symptoms and mental disorders may be more prevalent in certain cultures, but it should be noted that this is largely determined by whether that particular symptom(s) or disorder(s) is rooted more in genetic or social factors. For example, the prevalence of schizophrenia is consistent throughout the world cross-culturally, but pain, anxiety, depression, PTSD, and suicide rates are more attributed to cultural factors (Chiao, Li, Turner, Lee-Tauler, & Pringle, 2017; US Department of Health and Human Services, 2001). It should be noted that the elements of "self" (i.e., a set of beliefs, attitudes, thoughts, values, and emotions) in different cultures are central to the understanding of the conceptions of mental health, psychological well-being, and subsequent stigma as well as treatment-seeking (Kohrt & Harper, 2008; Liddell & Jobson, 2016), which are also significantly impacted by cultures (see Section "Cultural differences in social behavior and self-related processes").

Cross-cultural health model for prevention and treatment of disorders

In the same vein, research has shown that behavior change is also related to cultural differences in response to behavioral interventions or treatments (Jimenez et al., 2013, 2012; Kramer et al., 2002). Since differences in perception, attention, and subsequent cognitive/affective processing exist between Western and Eastern cultures (Markus & Conner, 2014; Nisbett, 2004), these data suggest that more work is necessary to understand how prevention and treatment should be tailored with respect to different cultures for effective changes (e.g., development of individualized culture-specific prevention or treatment programs). Is there a cross-cultural health model for the prevention and treatment of disorders?

Although cultural differences in attention, emotion, belief, feeling, and experience of symptoms and disorders have been shown, there is an increasing trend in global mental health of using integrative health approaches. Here, we discuss an Integrative Health Model (IHM) to address the concerns in global health and the prevention of disorders. The IHM has three components. (1) Conventional medicine which is taught in medical schools and widely adopted by mainstream healthcare services that focuses on disease treatment. In this framework people often believe symptom relief is equal to health and well-being and patients have the mindset that physicians should fix their problems and symptoms. Therefore many patients keep a passive attitude and mindset in response to treatment. (2) Complementary and alternative health care (CAH), which includes nonmainstream traditional health practices, but focuses more on prevention. These methods can translate into practices such as meditation, yoga, tai chi, herbal medicine, and acupuncture. Although the United States mainly emphasizes biomedical factors in the mental health and has adopted a medicalization model for mental health, about 38% of US adults and 12% of children are using some forms of CAH approaches that mainly focus on the self-regulation of body, mind, and behavior for health and well-being. The latest data from the National Health Interview Survey, published by National Institutes of Health and the National Center for Health Statistics, showed that more Americans meditated or practiced yoga in 2017 than in 2012 (Black, Barnes, Clarke, Stussman, & Nahin, 2018; Clarke, Barnes, Black, Stussman, & Nahin, 2018). CAH approaches such as mindfulness meditation is popular in research and in therapeutic applications (Tang et al., 2015). (3) Lifestyle and self-care, which mainly includes behavioral and nutritional approaches to promote health and wellness, and focuses more on prevention and healthy lifestyles. CAH and self-care approaches emphasize our own responsibilities, awareness, and care action to respond to health-related issues and prevent (or treat) mental disorders, but also encourage the combination of such approaches with conventional medicine techniques.

Conventional medicine as a mainstream healthcare approach has been widely applied in disease treatment; however, medical spending increases significantly (e.g., prescription drug spending is the fastest growing field in treatments). During the 1950s, about 2%−3% of the US gross domestic product (GDP) was for healthcare/medical spending; whereas since 2000 it increased to almost 20% of the GDP (Wilson, 2017). Unfortunately we have not received the highest quality of health care as we expected. People joke that sick care rather than health care applies under this system. Therefore a new healthcare model is urgently needed. IHM has been accepted by leading institutions such as Harvard, Stanford, and Columbia universities and has become a promising model for promoting global mental health and prevention of mental disorders in Western and Eastern cultures (Chase et al., 2018), but the implementation may be different cross-culturally. This raises an important question on how to develop the individualized and culturally adapted interventions or treatments to promote global mental health.

Clearly IHM's global implementation will be heavily affected by the beliefs, attitudes, emotions, and thinking processes of individuals in different cultures. Based on IHM, we propose two strategies to promote mental health cross-culturally: (1) understanding our experiences associated with symptoms and disorders in different cultures given that culture influences our experiences and also affects the way we describe our symptoms; and (2) selecting preventions or treatments following cultural differences given that people decide how to cope with mental disorders and seek treatments from psychiatrists, psychologists, physicians, or traditional healers based on the cultural effects of beliefs, attitudes, values, and norms (see Section "Cultural differences in emotion processes and mental health"). If we plan to develop a culture-specific prevention for mood disorders (i.e., using meditation for example), we should consider the cultural differences (e.g., global vs local, or holistic vs analytic information processing). Should we tailor the meditation program for people from Eastern and Western cultures using a more global or holistic strategy (i.e., open-monitoring meditation) that may promote intervention effects? Or should we take the opposite approach by using a more focal or analytic strategy (i.e., focus-based meditation) to adjust and balance the cultural influences? Should we also consider individual differences in learning, teaching, and practicing meditation within the same culture? These research gaps warrant further investigation. In our study we focused on cultural factors associated with mental health and summarized how cultures (e.g., Western and Eastern) may influence and shape our brain and behavior differently and thus affect our beliefs, attitudes, and experiences of symptoms and mental disorders. We discussed IHM to address the global medicalization issue and to promote global mental health and prevent mental disorders cross-culturally. The potential implications of cultural differences are the development of culturally adapted preventions or treatments for mental health promotion globally.

Cultural differences in genes and epigenetics

As discussed in Chapter 4, Genetic association with meditation learning and practice (outcomes), genes are one of the potential determinants of behavior and other psychological outcomes, while culture as the shared experience or environment also exerts influences on the expression of genes (epigenetics). Furthermore, cultural and genetic factors may interact to shape individuals' psychological and behavioral tendencies. We will consider collectivism versus individualism as an example to discuss gene–culture interaction. One study across 29 nations examined whether the prevalence of certain genotypes was associated with cultural tendencies such as collectivism or individualism. The results showed that more collectivistic nations had a greater historical prevalence of pathogens; the reason for this may be the higher frequencies of the short allele serotonin transporter-linked polymorphic region (5-HTTLPR) of the serotonin transporter gene SLC6A4 (Chiao & Blizinsky, 2010). Although the short allele 5-HTTLPR polymorphism is linked to high risk for anxiety and depression at the individual level, the national level showed an inverse correlation. Given that the cultural value of collectivism may buffer against pathogen prevalence, collectivism may also serve the adaptive function of reducing the risk, which may have led to genetic selection of the short allele of 5-HTTLPR in collectivistic cultures. In other words, people from Eastern collectivist cultures are more likely to have a gene that buffers them from depression than people from Western cultures (Chiao & Blizinsky, 2010; Fincher, Thornhill, & Murray, 2008).

A review article suggested that variations within the genes of serotonin (5-HTTLPR, MAOA) and opioid (OPRM1) are related to individual differences in social sensitivity such as collectivism and individualism. The relative frequency of variants in these genes is correlated with the degree of collectivism—individualism, suggesting that collectivism may have developed and persisted in populations with a high proportion of social sensitivity alleles because it was more compatible with such groups. Consistently, there was a correlation between the relative proportion of these alleles and lifetime prevalence of depression across nations. The collectivism—individualism status partially mediated the relationship between allele frequency and depression, suggesting that reduced depression in populations with a high proportion of social sensitivity alleles was due to greater collectivism (Way & Lieberman, 2010). These findings are in line with the results discussed in Chapter 4, which found cultural differences in social orientations were more significant for carriers of the DRD4 7- or 2-repeat alleles (e.g., higher levels of independence in Americans vs interdependence in Asians) than for noncarriers (Kitayama et al., 2014), together supporting the interaction between culture and genes.

Studies have suggested that the gene—culture interaction may occur in the brain. To integrate dynamic interactions between culture, behavior, brain, and gene, a culture-behavior-brain loop was proposed such that culture shapes the brain by contextualizing behavior, and the brain fits and modifies culture via behavioral influences. Genes provide a fundamental basis for, and interact with, the culture-behavior-brain loop at individual and population levels (Han & Ma, 2015). For instance, oxytocin-related genes may influence sensitivity toward social and emotional cues, but dopamine-related genes may affect sensitivity toward reward and punishment aspects of experiences. However, these genes may not influence sensitivity to cultural norms. The relative frequencies of these genes also vary within cultural groups (i.e., Chinese vs Americans), and some greater susceptibility alleles (e.g., the G allele of OXTR) are more common in Americans, whereas other greater susceptibility alleles (e.g., the short allele of 5-HTTLPR) are more common in Chinese. These findings may indicate that there is not a single gene for cultural conformity; instead, a complex set of genes may predispose us to be sensitive to different aspects of the cultural environment in different ways (Kim & Sasaki, 2014).

Cultural factors also influence how genetic predispositions manifest themselves in behavioral and psychological tendencies. In other words, cultures induce epigenetic changes. In Chapter 4, Genetic association with meditation learning and practice (outcomes), we discuss these questions and also point out meditation as a tailored experience that could influence gene expression——epigenetic changes through DNA methylation and/or histone modification. Moreover, epigenetic changes under cultures could induce intergenerational transmission——the transfer of traits from parents to offspring, including genetic and nongenetic influences—and thus may contribute to meditation acceptance, preference, learning, practice, and psychological and physiological outcomes in different cultures. For example, DRD4 7-repeat alleles are common in Western populations, whereas DRD4 2-repeat alleles are more prevalent among Asians. The DRD4 7-repeat allele was likely derived from a more ancient form ($\sim$45,000 years ago), whereas the 2-repeat allele appeared much later ($\sim$10,000 years ago), suggesting cultural differences between the West and East (Wang et al., 2004). These alleles may be under active selection pressures associated with migration, suggesting that DRD4 variants linked to altered dopamine signaling capacity could have coevolved with cultural forms of human adaptation. It is likely that behavioral traits associated with the 7-repeat or 2-repeat variant of DRD4 depend on what individuals learn from their experiences in a particular cultural context. For example, Americans or Chinese may learn through reward and reinforcement potentiated and are sensitized by the 7- or 2-repeat variant of DRD4 (Sasaki et al., 2013). Based on these findings, we should consider and apply different techniques and strategies to help meditators in different cultures. Meanwhile, meditation-induced epigenetic changes could also reshape our learning and practice of meditation and lead to different outcomes. For example, our series of RCTs using IBMT have suggested that meditation as a tailored experience is more effective and powerful (e.g., changes of behavior, physiology and brain) than culture, which often takes a longer time and across generations. How to learn and practice meditation correctly in different cultures is crucial to achieving benefits and positive outcomes.

Meditation–gene–culture interaction framework

As shown in Fig. 3.1, to explain our meditation ability and outcomes and how meditation influences our genes and behavior, we propose a meditation–gene–culture interaction framework including genetic factors, cultural system, meditation experience, learning, strategies, biological constraints, and behavior that may contribute to meditation learning, practice, and outcomes. It should be noted that meditation as a tailored experience may also shape the expression of genes and induces epigenetic changes that could transmit changes to intergeneration [see Chapter 4: Genetic association with meditation learning and practice (outcomes) for details].

In summary, we have discussed cultural differences in diverse domains such as perception, attention, cognition, emotion, social behavior, symptoms and disorders, genes and epigenetics, especially regarding how these

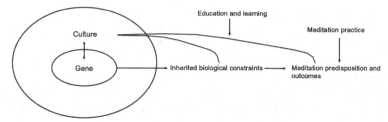

FIGURE 3.1 Meditation–gene–culture interaction framework.

macrolevel differences may impact the actual learning, practice, and outcomes of meditation. We focus on these differences between Eastern and Western cultures and their relation to meditation from behavioral, genetic, and neural perspectives. We also propose a cross-cultural health model and explore how to develop the culturally adapted preventions and interventions such as meditation to promote behavior change, health, and well-being. It is important to note that culture is a multifaceted concept encompassing a variety of factors that interact with one another and has far-reaching impacts, on not only behavior and health, but also other critical aspects of our daily lives. One important topic is personalized health promotion cross-culturally to increase participants' interest and engagement to maintain well-being and health. We propose that personalizing health promotion should follow individual's cultural beliefs, values, interests, and preferences to develop different types of interventions such as meditation cross-culturally. Given that research on cultural differences in meditation is still fairly limited, there is still a lot to be learnt regarding which cultural factors may play the greatest role in differentially affecting meditation practice, outcomes, and applications in health maintenance, prevention, and treatment of disorders across different cultures. Therefore future research needs to delineate the exact mechanisms and factors underlying the contribution of culture in shaping the interpretations and experiences of feelings, symptoms, and disorders.

Practice tips

1. It is crucial to train somatic or interoceptive awareness through regulating and optimizing body postures, which could help and facilitate the meditative state. In IBMT research, we use both bodifulness and mindfulness to improve somatic or interoceptive awareness and achieve great effectiveness. This combined training approach may help meditators in different cultures to take inner feeling and sensation as a target and use less effort or achieve effortlessness.
2. Analytic and logical versus holistic and contextual styles of mental processes could influence meditation decision, learning, and practice in Westerners and Easterners (e.g., Americans vs Chinese). As of now, there

is no data that directly supports the notion that Americans are good at focus-based meditation. However, our preliminary results suggested that Americans benefit from a context-based strategy during meditation (e.g., see both target and context simultaneously).

3. People from Western and Eastern cultures benefit more from open-focus practice than narrow-focus practice, as narrow focus involves too much effort and control that interferes with the practice. Our IBMT experience suggests to use percentage effort ratio to achieve promising results, for example, 20:80 or 30:70 for effort on the target to context ratio.

4. Cultures affect the self and self-related processes, meditation also involves processes such as self-reconstructing, deconstructing, or transcendence experiences. Self-inquiry based on Zen helps self-transformation processing that may fit into different cultures.

5. White bear phenomena helps meditation practice in different cultures. Our IBMT experience always suggests not to suppress or control thoughts, feelings, or experiences during meditation because avoidance or suppression (of a thought) often leads to thinking about the very thought one hopes to suppress. In other words, what we resist persists. Thus, attempts to suppress thoughts (e.g., using substances) may actually lead to increases in substance use.

References

Bar-Haim, Y., Lamy, D., Pergamin, L., Bakermans-Kranenburg, M. J., & van IJzendoorn, M. H. (2007). Threat-related attentional bias in anxious and nonanxious individuals: A meta-analytic study. *Psychological Bulletin, 133*(1), 1−24.

Black, L. I., Barnes, P. M., Clarke, T. C., Stussman, B. J., & Nahin, R. L. (2018). *Use of yoga, meditation, and chiropractors among U.S. children aged 4−17 years.* Hyattsville, MD: National Center for Health Statistics, NCHS Data Brief, 324.

Boduroglu, A., Shah, P., & Nisbett, R. E. (2009). Cultural differences in allocation of attention in visual information processing. *Journal of Cross-Cultural Psychology, 40*, 349−360.

Browning, M., Holmes, E. A., & Harmer, C. J. (2010). The modification of attentional bias to emotional information: A review of the techniques, mechanisms, and relevance to emotional disorders. *Cognitive, Affective and Behavioral Neuroscience, 10*(1), 8−20.

Cavanna, A. E., & Trimble, M. R. (2006). The precuneus: A review of its functional anatomy and behavioural correlates. *Brain, 129*, 564−583.

Chase, L. E., Sapkota, R. P., Crafa, D., & Kirmayer, L. J. (2018). Culture and mental health in Nepal: An interdisciplinary scoping review. *Global Mental Health (Cambridge), 5*, e36.

Chiao, J. Y., & Blizinsky, K. D. (2010). Culture-Cgene coevolution of individualism−collectivism and the serotonin transporter gene (5-HTTLPR). *Proceedings of the Royal Society B, 277*, 529−537.

Chiao, J. Y., Cheon, B. K., Pornpattananangkul, N., Mrazek, A. J., & Blizinsky, K. D. (2013). Cultural neuroscience: Progress and promise. *Psychological Inquiry, 24*, 1−19.

Chiao, J. Y., Harada, T., Komeda, H., Li, Z., Mano, Y., Saito, D., ... Iidaka, T. (2009). Neural basis of individualistic and collectivistic views of self. *Human Brain Mapping, 30*(9), 2813−2820.

Chiao, J. Y., Li, S.-C., Seligman, R., & Turner, R. (Eds.), (2016). *Oxford handbook of cultural neuroscience*. New York, NY: Oxford University Press.

Chiao, J. Y., Li, S. C., Turner, R., Lee-Tauler, S. Y., & Pringle, B. A. (2017). Cultural neuroscience and global mental health: Addressing grand challenges. *Culture and Brain, 5*(1), 4−13.

Chua, H. F., Boland, J. E., & Nisbett, R. E. (2005). Cultural variation in eye movements during scene perception. *Proceedings of National Academy Sciences of United States of America, 102*(35), 12629−12633.

Clarke, T. C., Barnes, P. M., Black, L. I., Stussman, B. J., & Nahin, R. L. (2018). *Use of yoga, meditation, and chiropractors among U.S. adults aged 18 and over*. Hyattsville, MD: National Center for Health Statistics, NCHS Data Brief, 325.

Craig, A. D. (2015). *How do you feel? An interoceptive moment with your neurobiological self*. Princeton, NJ: Princeton University Press.

Critchley, H. D., Mathias, C. J., Josephs, O., O'Doherty, J., Zanini, S., Dewar, B. K., et al. (2003). Human cingulate cortex and autonomic control: converging neuroimaging and, clinical evidence. *Brain, 126*, 2139−2152.

Daniel, L. S., & Donna Rose, A. (2007). The cognitive neuroscience of constructive memory: Remembering the past and imagining the future. *Philosophical Transactions of the Royal Society B: Biological Sciences, 362*, 773−786.

Di Lernia, D., Serino, S., & Riva, G. (2016). Pain in the body. Altered interoception in chronic pain conditions: A systematic review. *Neuroscience and Biobehavioral Reviews, 71*, 328−341.

Fincher, C. L., Thornhill, R., & Murray, D. R. (2008). SchallerM. Pathogen prevalence predicts human cross-cultural variability in individualism/collectivism. *Proceedings of the Royal Society B, 275*, 1279−1285.

Gajewski, P. D., Stoerig, P., & Falkenstein, M. (2008). ERP—correlates of response selection in a response conflict paradigm. *Brain Research, 1189*, 127−134.

Goh, J. O., & Park, D. C. (2009). Culture sculpts the perceptual brain. *Progress in Brain Research, 178*, 95−111.

Goldin, P. R., McRae, K., Ramel, W., & Gross, J. J. (2008). The neural bases of emotion regulation: Reappraisal and suppression of negative emotion. *Biological Psychiatry, 63*, 577−586.

Greene, J. D., Nystrom, L. E., Engell, A. D., Darley, J. M., & Cohen, J. D. (2004). The neural bases of cognitive conflict and control in moral judgment. *Neuron, 44*, 389−400.

Harris, K., Spiegelhalder, K., Espie, C. A., MacMahon, K. M., Woods, H. C., & Kyle, S. D. (2015). Sleep-related attentional bias in insomnia: A state-of-the-science review. *Clinical Psychology Review, 42*, 16−27.

Hedden, T., Ketay, S., Aron, A., Markus, H. R., & Gabrieli, J. D. E. (2008). Cultural influences on neural substrates of attentional control. *Psychological Science, 19*, 12−17.

Han, S., & Ma, Y. (2014). Cultural differences in human brain activity: A quantitative meta-analysis. *Neuroimage, 99*, 293−300.

Han, S., & Ma, Y. (2015). A culture-behavior-brain loop model of human development. *Trends Cogn Sci, 19*(11), 666−676.

Hunt, J., Sullivan, G., Chavira, D. A., Stein, M. B., Craske, M. G., Golinelli, D., Roy-Byrne, P. P., & Sherbourne, C. D. (2013). Race and beliefs about mental health treatment among anxious primary care patients. *J Nerv Ment Dis, 201*(3), 188−195.

Hughes, A. M., Hirsch, C. R., Nikolaus, S., Chalder, T., Knoop, H., & Moss-Morris, R. (2018). Cross-cultural study of information processing biases in chronic fatigue syndrome: comparison of Dutch and UK chronic fatigue patients. *International Journal of Behavioral Medicine, 25*(1), 49−54.

Jimenez, D. E., Bartels, S. J., Cardenas, V., & Alegría, M. (2013). Stigmatizing attitudes toward mental illness among racial/ethnic older adults in primary care. *International Journal of Geriatric Psychiatry*, *28*(10), 1061−1068.

Jimenez, D. E., Bartels, S. J., Cardenas, V., Dhaliwal, S. S., & Alegría, M. (2012). Cultural beliefs and mental health treatment preferences of ethnically diverse older adult consumers in primary care. *The American Journal of Geriatric Psychiatry*, *20*(6), 533−542.

Kim, H. S., & Sasaki, J. Y. (2014). Cultural neuroscience: Biology of the mind in cultural contexts. *Annual Review of Psychology*, *65*, 487−514.

Kitayama, S., King, A., Yoon, C., Tompson, S., Huff, S., & Liberzon, I. (2014). The Dopamine D4 Receptor Gene (DRD4) moderates cultural difference in independent versus interdependent social orientation. *Psychological Science*, *25*(6), 1169−1177.

Kitayama, S., & Park, J. (2010). Cultural neuroscience of the self: Understanding the social grounding of the brain. *Social Cognitive and Affective Neuroscience*, *5*, 111−129.

Kohrt, B. A., & Harper, I. (2008). Navigating diagnoses: Understanding mind-body relations, mental health, and stigma in Nepal. *Culture, Medicine and Psychiatry*, *32*, 462−491.

Kramer, E. J., Kwong, K., Lee, E., & Chung, H. (2002). Cultural factors influencing the mental health of Asian Americans. *The Western Journal of Medicine*, *176*(4), 227−231.

Li, J. B., Vazsonyi, A. T., & Dou, K. (2018). Is individualism-collectivism associated with self-control? Evidence from Chinese and U.S. samples. *PLoS ONE*, *13*(12), e0208541.

Liddell, B. J., & Jobson, L. (2016). The impact of cultural differences in self-representation on the neural substrates of posttraumatic stress disorder. *European Journal of Psychotraumatology*, *7*, 30464.

Lutz, J., Brühl, A. B., Scheerer, H., Jäncke, L., & Herwig, U. (2016). Neural correlates of mindful self-awareness in mindfulness meditators and meditation-naïve subjects revisited. *Biological Psychology*, *119*, 21−30.

Ma-Kellams, C. (2014). Cross-cultural differences in somatic awareness and interoceptive accuracy: A review of the literature and directions for future research. *Frontiers in Psychology*, *5*, 1379.

Markus, H. R., & Conner, A. C. (2014). *Clash! How to thrive in a multicultural world*. New York: Penguin (Hudson Street Press).

Markus, H. R., & Kitayama, S. (1991). Culture and the self: Implications for cognition, emotion, and motivation. *Psychological Review*, *98*(2), 224−253.

Matsumoto, D., Yoo, S. H., & Nakagawa, S. (2008). Multinational study of cultural display rules. Culture, emotion regulation, and adjustment. *Journal of Personality and Social Psychology*, *94*, 925−937.

Mrazek, A. J., Chiao, J. Y., Blizinsky, K. D., Lun, J., & Gelfand, M. J. (2013). The role of culture−gene coevolution in morality judgment: Examining the interplay between tightness−looseness and allelic variation of the serotonin transporter gene. *Culture and Brain*, *1* (2-4), 100−117.

Nisbett, R. E. (2004). *The geography of thought: How Asians and Westerners think differently and why*. New York: Free Press.

Nisbett, R. E., & Miyamoto, Y. (2005). The influence of culture: Holistic versus analytic perception. *Trends in Cognitive Sciences*, *9*(10), 467−473.

Nisbett, R. E., Peng, K., Choi, I., & Norenzayan, A. (2001). Culture and systems of thought: Holistic versus analytic cognition. *Psychological Review*, *108*, 291−310.

Ohira, H., Nomura, M., Ichikawa, N., Isowa, T., Iidaka, T., et al. (2006). Association of neural and physiological responses during voluntary emotion suppression. *NeuroImage*, *29*, 721−733.

Oyserman, D., & Lee, S. W. (2008). Does culture influence what and how we think? Effects of priming individualism and collectivism. *Psychological Bulletin*, *134*, 311–342.

Prehn, K., Wartenburger, I., Mériau, K., Scheibe, C., Goodenough, O. R., Villringer, A., & Heekeren, H. R. (2008). Individual differences in moral judgment competence influence neural correlates of socio-normative judgments. *Social Cognitive Affective Neuroscience*, *3*, 33–46.

Rozin, P. (2003). Five potential principles for understanding cultural differences in relation to individual differences. *Journal of Research in Personality*, *37*, 273–283.

Rusu, A. C., Gajsar, H., Schlüter, M. C., & Bremer, Y. I. (2019). Cognitive biases towards pain: Implications for a neurocognitive processing perspective in chronic pain and its interaction with depression. *The Clinical Journal of Pain*, *35*(3), 252–260.

Sasaki, J. Y., Kim, H. S., Mojaverian, T., Kelley, L. D., Park, I. Y., & Janušonis, S. (2013). Religion priming differentially increases prosocial behavior among variants of the dopamine D4 receptor (DRD4) gene. *Social Cognitive and Affective Neuroscience*, *8*, 209–215.

Tan, L. H., Xu, M., Chang, C. Q., & Siok, W. T. (2013). China's language input system in the digital age affects children's reading development. *Proceedings of the National Academy of Sciences of the United States of America*, *110*(3), 1119–1123.

Tang, Y. Y., Holzel, B. K., & Posner, M. I. (2015). The neuroscience of mindfulness meditation. *Nature Reviews Neuroscience*, *16*, 213–225.

Tang, Y. Y. (2009). *Exploring the brain, optimizing the life*. Beijing: Science Press.

Tang, Y. Y., Ma, Y., Wang, J., Fan, Y., Feng, S., Lu, Q., & Posner, M. I. (2007). Short term meditation training improves attention and self regulation. *Proceedings of the National Academy of Sciences, USA*, *104*(43), 17152–17156.

Tang, Y. Y. (2017). *The neuroscience of mindfulness meditation: How the body and mind work together to change our behavior?* London: Springer Nature.

Tang, Y. Y., & Liu, Y. (2009). Numbers in the cultural brain. *Progress in Brain Research*, *178*, 151–157.

Tang, Y. Y., & Tang, R. (2014). Ventral-subgenual anterior cingulate cortex and self-transcendence. *Frontiers in Psychology*, *4*, 1000.

Tang, Y. Y., & Tang, R. (2015a). Rethinking the future directions of mindfulness field. *Psychological Inquiry*, *26*(4), 368–372.

Tang, Y. Y., & Tang, R. (2015b). Mindfulness: Mechanism and application. In A. W. Toga (Ed.), *Brain mapping: An encyclopedic reference* (3, pp. 59–64). Academic Press, Elsevier.

Tang, Y. Y., Tang, R., & Gross, J. J. (2019). Promoting emotional well-being through an evidence-based mindfulness training program. *Frontiers in Human Neuroscience*, *13*, 237.

Tang, Y. Y., Zhang, W., Chen, K., Feng, S., Ji, Y., Shen, J., ... Liu, Y. (2006). Arithmetic processing in the brain shaped by cultures. *Proceedings of the National Academy of Sciences of United States of America*, *103*(28), 10775–10780.

Tang, Y. Y., Tang, R., & Posner, M. I. (2013). Brief meditation training induces smoking reduction. *Proceedings of the National Academy of Sciences, USA*, *110*(34), 13971–13975.

Tang, Y. Y., Lu, Q., Fan, M., Yang, Y., & Posner, M. I. (2012). Mechanisms of white matter changes induced by meditation. *Proceedings of the National Academy of Sciences, USA*, *109*(26), 10570–10574.

Tang, Y. Y., Lu, Q., Geng, X., Stein, E. A., Yang, Y., & Posner, M. I. (2010). Short-term meditation induces white matter changes in the anterior cingulate. *Proceedings of the National Academy of Sciences, USA*, *107*(35), 15649–15652.

Tompson, S. H., Huff, S. T., Yoon, C., King, A., Liberzon, I., & Kitayama, S. (2018). The dopamine D4 receptor gene (DRD4) modulates cultural variation in emotional experience. *Culture and Brain*, 6(2), 118−129.

U.S. Department of Health and Human Services. (2001). *Mental health: Culture, race, and ethnicity—A supplement to mental health: A report of the surgeon general.* Rockville, MD: U.S. Department of Health and Human Services, Substance Abuse and Mental Health Services Administration, Center for Mental Health Services.

Wang, E., Ding, Y.-C., Flodman, P., Kidd, J. R., Kidd, K. K., Grady, D. L., . . . Moyzis, R. K. (2004). The genetic architecture of selection at the human dopamine receptor D4 (DRD4) gene locus. *American Journal of Human Genetics*, 74, 931−944.

Wang, Y., Deng, Y., Sui, D., & Tang, Y. Y. (2014). Neural correlates of cultural differences in moral decision making: A combined ERP and sLORETA study. *Neuroreport*, 25, 110−116.

Way, B. M., & Lieberman, M. D. (2010). Is there a genetic contribution to cultural differences? Collectivism, individualism and genetic markers of social sensitivity. *Social Cognitive and Affective Neuroscience*, 5, 203−211.

Wilson, K. (2017). Health Care Costs 101: Spending Rose with More Coverage and Care, https://www.chcf.org/publication/health-care-costs-101-spending-rose-with-more-coverage-and-care/

Chapter 4

Genetic association with meditation learning and practice (outcomes)

What are genes and alleles?

A gene is a unit of hereditary information and is composed of DNA that codes genetic information for the transmission of heritable traits. Alleles are genetic sequences, but variant forms of genes. A given gene may have multiple different alleles, and alleles could result in different phenotypes in the observable genetic expression. Some alleles are dominant in that they override the expression of other alleles, or sometimes multiple alleles can be expressed at the same time in a codominant fashion. For instance, in the human ABO blood group system, if a person has type AB blood, she or he has one allele for A and one allele for B, indicating that both alleles are expressed. Different genes and alleles can affect our brain, physiology, and behavior, leading to observable and even latent individual differences (Champagne & Mashoodh, 2012).

How do genes and alleles influence behaviors?

Growing evidence indicates that genetic variation and alleles contribute to individual differences in behavior and the brain and, therefore, have the potential to influence meditation practices. Temperamental or personality traits such as effortful control or inhibitory control (one form of self-control capacity) or sensation-seeking has important implications for children's development. Although genetic factors and parenting may influence temperamental traits, few studies have examined the interplay between the two in predicting its development. For example, previous research has shown that the dopamine receptor D4 gene (DRD4) 7-repeat allele interacts with environmental factors such as learning and parenting in children and adults, which then influence children's behaviors such as risk taking and effortful or inhibitory control (Smith, Kryski, Sheikh, Singh, & Hayden, 2013; Smith et al., 2012).

The Neuroscience of Meditation. DOI: https://doi.org/10.1016/B978-0-12-818266-6.00005-8

In toddlers, lower-quality parenting in combination with the 7-repeat allele of the DRD4 gene is associated with greater parent-reported temperamental trait in sensation-seeking, but not in effortful control (Sheese, Voelker, Rothbart, & Posner, 2007). A follow-up assessment with the same sample of children showing that parenting quality interacts with the presence of the DRD4 7-repeat allele to predict effortful control in 3 to 4-year-old children. These findings may reflect the increased role of the effortful control in executive attention network in older children and adults (Sheese, Rothbart, Voelker, & Posner, 2012). However, due to the small sample size ($N = 52$), these findings should be treated with caution and considered as preliminary evidence until they are replicated in an independent and large sample. A later study with a large sample of 3-year-old children ($N = 409$) used the same assessments to investigate whether the associations between parenting and children's effortful control and inhibitory control were moderated by 7-repeat alleles. Results showed that when the quality of parenting is lower, children with at least one 7-repeat allele display lower inhibitory control than those without this allele, suggesting 7-repeat alleles moderate the association between parenting quality and inhibitory control and such genetic polymorphisms may increase the vulnerability to environmental or contextual influences (Smith et al., 2013).

Does dopamine receptor D4 moderate cultural differences in learning, emotional experience, and social behavior?

Culture is an important aspect of the environment or experience (e.g., a "social intervention"). The carriers of DRD4 alleles may be more likely to show culturally typical response patterns than noncarriers. For instance, white Americans report experiencing positive emotions more than negative emotions, but this pattern is atypical for Asians. One study tested whether the positivity bias in emotional experience is moderated by DRD4 and culture in white Americans and Asians. Results showed a significant culture × DRD4 interaction for emotional experience, such that Asian carriers of the 7- or 2-repeat alleles reported experiencing greater emotional balance (i.e., weaker positivity bias) than noncarriers. However, the pattern was reversed in white Americans such that the positivity bias was stronger among the carriers (Tompson et al., 2018).

Further studies also investigated whether DRD4 plays a role in modulating cultural learning and social behavior such as independent versus interdependent social orientation. For example, previous research suggests that white Americans are more independent (less interdependent) than Asian-born Asians and that cultural groups vary on social orientation. Results showed that cultural difference was significantly more pronounced for carriers of the 7- or 2-repeat alleles (higher levels of culturally dominant social orientations) than for noncarriers (Kitayama et al., 2014). Taken together, these results suggest that people who carry a 7- or 2-repeat allele of DRD4 genes are more sensitive to environmental or experiential influences than those who do not carry this allele.

Does dopamine receptor D4 affect training or intervention?

Based on previous research, it is reasonable to expect that DRD4 7-repeat alleles would affect training or intervention. In a randomized controlled trial (RCT) with a large sample of 5-year-old children ($N = 508$), 257 children showed a delay in literacy skills as measured by a national standard literacy test in the Netherland. The study examined whether children with DRD4 7-repeat alleles were more responsive to educational interventions than noncarriers. Children were randomly assigned to three groups, a control condition and two computerized interventions tailored to the literacy needs—Living Letters for alphabetic knowledge and Living Books for text comprehension. For carriers of DRD4 7-repeat alleles (about one-third of the delayed group), the Living Books intervention was effective, but it did not affect noncarriers, suggesting effects of differential susceptibility. The carriers also benefited more from the Living Letters intervention than the noncarriers, suggesting the DRD4 7-repeat alleles can also influence training or intervention effects (Plak, Kegel, & Bus, 2015). Similar results were reported that carriers of DRD4 7-repeat alleles benefited more from interventions that seek to decrease externalizing behavior and reduce substance use (Bakermans-Kranenburg, Van IJzendoorn, Pijlman, Mesman, & Juffer, 2008; Brody et al., 2014). These findings indicate that we are differentially responsive to intervention effects depending on genetic variations. Why is the DRD4 involved in emotional experience, social behavior, and intervention? A brief answer is that the dopamine receptor is responsible for neuronal signaling in the mesolimbic system of the brain, an area involved in emotion and complex behavior. Meditation, a form of intervention, is thus very likely moderated by DRD4 7-repeat alleles, and the carriers of these gene alleles could be more responsive to meditation.

Are there other genotypes that affect behavior and intervention?

No single candidate gene or gene allele could explain all the variability in intervention responsiveness. Therefore investigations of interactions among these gene alleles or polymorphisms are important to provide a more comprehensive account of the genetic underpinnings of certain behaviors. In addition to DRD4, are there other genotypes affecting behavior and intervention?

Neurotransmitter serotonin, or 5-hydroxytryptamine (5-HT), helps regulate diverse functions including mood, cognition, social behavior, appetite, digestion, and sleep. Serotonin transporters transport the serotonin from synaptic spaces into presynaptic neurons and play an important role in psychiatric disorders. The serotonin transporter (5-HTT) gene promoter variant is one of the major factors contributing to the etiology of many psychiatric disorders such as posttraumatic stress disorder and depression-susceptibility (Kuzelova, Ptacek, & Macek, 2010). A longitudinal study investigated why stressful experiences lead to depression in some people, but not in others. Results suggested a polymorphism in the promoter region of the 5-HTT gene moderated the influence of stressful life events on depression. In particular, individuals with one or two copies of the 5-HTT short allele showed more depressive symptoms and suicidality related to stress than individuals with the long allele (Caspi et al., 2003). The study indicated that how we respond to the environment is partly moderated by our genetic alleles.

A further study examined how genetic and experiential factors interact with a family-based training exercise during development in preschool children coming from lower socioeconomic status (SES) backgrounds (Isbell et al., 2017). Findings indicated that the family-based training could modify the associations between genotype serotonin transporter-linked polymorphic region (5-HTTLPR) of the serotonin transporter gene SLC6A4 and the individual differences in selective attention. In particular, compared to the 5-HTTLPR short-allele carriers, the 5-HTTLPR long-allele carriers may denote a risk factor for selective attention in preschool-age children from lower SES backgrounds. Why is 5-HTTLPR involved in this study? The prefrontal cortex (PFC) is densely populated with serotonin receptors and transporter sites, and the serotonergic system contributes to the development, neuroplasticity, and functioning of the PFC, such as selective attention (Puig & Gulledge, 2011).

It is critical to examine how 5-HTTLPR interacts with other candidate polymorphisms. One study indicated both DRD4 and 5-HTTLPR genetic variants influence intervention effects on alcohol use of early adolescents by maternal involvement (Cleveland et al., 2015). Prior studies reported gene × intervention effects for polymorphisms of various candidate genes, including (but not limited to) the DRD4, dopamine-active transporter, and brain-derived neurotrophic factor (BDNF), suggesting multiple genes and their interaction can moderate intervention effects (Albert et al., 2015; Isbell et al., 2017). It should be noted that gene × intervention may depend on a multitude of factors, such as the candidate genes, types of interventions, the characteristics of interventions, age, gender, and personality.

For example, in Chapter 2, Personality and meditation, we discuss how personality traits and other individual characteristics influence meditation practices in terms of practice engagement, frequency, and the effectiveness of achieving the desirable outcomes. Some of our studies have further shown that personality and mood can predict individual differences in the improvement of creative performance following mindfulness meditation, such as integrative body-mind training (IBMT) (Ding, Tang, Deng, Tang, & Posner, 2015;

Ding, Tang, Tang, & Posner, 2014). Given that potential genes also interact with personality traits, it is highly possible that gene × personality may moderate the quality of meditation learning and practice.

Genetic polymorphisms and treatment response

As we know, people differ in treatment response. One reason could be the genetic influence due to different genetic alleles or polymorphisms presented among us. For instance, one study investigated the heterogeneity of therapeutic responses to brain stimulation techniques. More than 20 genetic variants within at least 10 genes (e.g., BDNF, COMT, DRD2, and 5-HHT) were examined and the results suggested the polymorphism of human BDNF gene——Val66Me moderated treatment response. Since Val66Me plays an important role in neuroplasticity, these results may suggest that gene susceptibility to treatment response depends on treatment types. For instance, treatment responses to brain stimulation and meditation may be moderated by different genes and underlying mechanisms (Saghazadeh, Esfahani, & Rezaei, 2016). Similarly, to explore genes related to major depressive disorder (MDD) and antidepressant treatment response, a study found differences in the expression of several genes before and after various antidepressant treatments in MDD. These altered genes (e.g., CXCL8) in MDD are mainly involved in the immune response and inflammation and can predict which patients respond well to antidepressants (Woo, Lim, Myung, Kim, & Lee, 2018). A recent study assessed the potential relationship between genetic polymorphisms and drug treatment response in Alzheimer's disease (AD) and found that genetic polymorphisms of ABCA1, ApoE3, CHRNA7, and ESR1 genes seem to have strong correlations with the treatment response in AD (Sumirtanurdin, Thalib, Cantona, & Abdulah, 2019). These findings consistently suggest genes, disorders, and treatment types may interact to moderate treatment responses.

Only a few pilot studies have explored the possible genetic association with mind−body intervention responses. Growing evidence suggests that the 5-HTTLPR variant of the serotonin transporter gene (SLC6A4) and MTHFR 677C > T polymorphisms are linked to MDD and antidepressant treatment response. In a 12-week RCT, MDD patients ($N = 178$) were randomized to receive yoga-based lifestyle intervention or drug therapy. The goal was to examine the impact of yoga on MDD patients who had 5-HTTLPR and MTHFR 677C > T polymorphisms. Results showed that yoga provided remission in MDD patients but neither 5-HTTLPR or MTHFR 677C > T polymorphisms showed any influence on remission (Tolahunase, Sagar, Faiq, & Dada, 2018). Another study applied mindfulness-based stress reduction (MBSR) in veterans with posttraumatic stress disorder (PTSD) (the ratio of responders to nonresponders was 11:11) to identify potential epigenetic factors relevant for treatment response. Given that serotonin signaling and hypothalamic–pituitary–adrenal (HPA) axis functioning are often affected by PTSD, the study targeted changes in these two molecular pathways following PTSD treatments. Results indicated that responders had a decrease in FKBP5 methylation, but nonresponders had an increase in FKBP5, suggesting that effective meditation intervention may be associated with stress-related pathways at the molecular level. These preliminary findings suggest that DNA methylation signatures within FKBP5 are potential indicators of meditation responsiveness in PTSD patients, but warrant further replication in larger sample sizes (Bishop et al., 2018).

Given the limited evidence, we could not draw any conclusion from these preliminary results, but they raise an important question on how to help non-responders to better practice and benefit from meditation. Meditation entails a form of mental-state training which targets attention and self-control networks in the brain (Tang & Posner, 2009). Any technique that strengthens attention and self-control networks may help and facilitate meditation states. For example, a pilot study suggested that brain stimulation potentially enhanced the effects of mindfulness meditation (Badran et al., 2017). Combining meditation with other methods such as neurofeedback or brain stimulation may also increase meditation effects on our targeted outcomes. The key to such approaches is that they should be able to synergistically strengthen attention and self-control networks to facilitate meditation practices (Tang, 2017; Tang, Tang, & Gross, 2019; Tang, Tang, Rothbart, & Posner, 2019).

What we have learned from genes and alleles thus far is that we should consider several candidate genes such as DRD4, 5-HTT and 5-HTTLPR, which may predispose some people (e.g., carriers of DRD4 7-repeat alleles, 5-HTT short allele, or/and 5-HTTLPR short-allele) to be more sensitive to, and benefit from, meditation learning and practice. Furthermore, the gene x meditation interaction could potentially influence our brain functioning, plasticity, and behavior changes. Of note, there might be meditation sensitive gene(s) that we have not found yet and the carriers of these genes/alleles may be particularly suitable for meditation practices. For instance, some people could participate in and finish long-term meditation retreats with more than 8 h/day, while others could not tolerate meditation even for half an hour per day, suggesting pre-existing differences and gene variations may play an important role in this difference. It is also plausible that carriers with certain genes/alleles are more responsive to certain meditation techniques than other alternatives. However, currently not much is known about the precise mechanisms of how genes interact with the mind and body to affect behavior such as meditation practice and responsiveness to specific techniques. In the future, with more evidence-based research, we will be able to describe how genes may come into play when individuals learn to practice meditation. Luckily, it is increasingly possible to apply new genetic findings to guide people on how to select the most appropriate intervention and benefit more from the intervention.

Practice tips

1. For carriers of DRD4 or 5-HTTLPR, we could encourage them to experience and practice different meditation techniques to select the most effective option.
2. For noncarriers, we could use different strategy, for example, encourage them to experience one meditation technique, but also remind them to be patient and take time to gradually build the learning curve. Of note, we should be open-minded and flexible to noncarriers since other factors such as belief, attitude, personality, and lifestyle could also affect meditation learning and practice outcomes.
3. To help nonresponders, we could combine other methods such as brain stimulation or neurofeedback to facilitate meditation states and practices. For instance, we could first use brain stimulation or neurofeedback to change their baseline brain state to be ready for the subsequent meditation practice. The length, frequency, target areas, and dosage of the stimulation are critical for it to be effective.
4. We usually focus more on meditation techniques, but having a correct attitude and understanding of meditation are crucial because these factors can also decide where our attention, mental effort, and practice will be directed during meditation. If we make sure that participants are appropriately guided in terms of attitude and understanding, then we may be able to induce epigenetic changes following meditation in nonresponders.

Epigenetics: how do experience or environment shapes our genes?

Our genes affect our experience and behavior such as emotion, cognition, learning, and practice, but to what extent might experience or environment also alter gene activity and expression and consequently influence our brain function and behavior? Epigenetics involves heritable phenotype changes such as affecting gene activity and expression through DNA methylation and histone modification, but it does not involve alterations in the DNA sequences (Francis, 2012). Such effects on cellular and physiological traits may result from lifespan experiences (or environmental factors) such as early life stress, addiction, fear conditioning, depression and anxiety, development, or intervention. For example, animal studies have shown that certain conditional fears can be inherited from either parent. In the experiment, mice were conditioned to fear a strong scent (acetophenone) by accompanying the smell with an electric shock. Consequently, this fear could be passed down to the mice offspring and these epigenetic changes could last up to two generations without reintroducing the shock. Even though the offspring never experienced the shock themselves, the mice still displayed a fear for the acetophenone scent, suggesting that the mice inherited the fear epigenetically (Szyf, 2014).

Addiction or substance use has been shown to affect our brain function, structure, and behavior (Tang, Posner, Rothbart, & Volkow, 2015). Long-term brain changes after chronic exposure to drug abuse suggest that alterations in gene regulation contribute to the addictive phenotype through multiple mechanisms——drugs alter the transcriptional potential of genes from transcriptional machinery (mobilization or repression) to epigenetics. Similar to drug-exposure experience, poverty, nutrition, socioeconomic status, psychosocial factors, and other experiences are also associated with DNA methylation of epigenetics (Coker, Gunier, Huen, Holland, & Eskenazi, 2018; Lam et al., 2012). These studies raise an important nature versus nurture question.

Nature versus nurture

Behavioral epigenetics is a new research field that examines how epigenetics affects behavior and how nurture shapes nature (Miller, 2010). Nature refers to biological heredity and nurture refers to what that occurs to us during the course of life, such as diet and nutrition, stress, emotional and social experience, and exposure to toxins. Behavioral epigenetics aims to understand how gene expression is influenced by experiences and environment to produce individual differences in cognition, emotion, personality, behavior, and mental health. These epigenetic changes influence the growth of neurons in developing brains and modify activity of the neurons in adult brains. Therefore the epigenetic changes on neuron structure and function have a marked influence on behavior (Moore, 2015; Zhang & Meaney, 2010).

For example, even short-term substance abuse produces long-lasting epigenetic changes in the brain reward system of rodents. These epigenetic changes modify gene expression, which, in turn, increases the vulnerability of repeated substance overdose in the future (Renthal & Nestler, 2008). Epigenetic modifications are also associated with diverse and different disease conditions such as cancer, eating disorders and obesity, diabetes, schizophrenia, bipolar disorder, and depression (Bronner & Helmtrud, 2011).

Based on a growing body of evidence that life experiences and environment alter the way by which genes are marked with DNA methylation in the brain and peripheral tissues, we hypothesize that changes in DNA methylation and other epigenetic markers could generate stable phenotypes associated with brain changes and diverse behavioral patterns such as stress response, emotion regulation, and risk taking (Szyf, Tang, Hill, & Musci, 2016). Although the data are still sparse, epigenetic studies have illustrated that experiences and environment can influence genes, and that epigenetic changes occur over time in response to experiences and environment. Since DNA methylation is potentially preventable and reversible, it is possible to develop epigenetically targeted interventions (Szyf et al., 2016). The research findings also suggest a potential application——how to select and tailor certain experience such as meditation to each individual and design epigenetically informed meditation techniques to voluntarily shape our biological heredity, thus changing our behavior and improving health, well-being, and quality of life?

Meditation and epigenetics

Meditation as an experience could influence gene expression. However, so far, little is known about epigenetic responses to meditation practice. Some studies have explored the potential association between genomic and clinical effects of mind—body interventions such as relaxation response (RR) and meditation in healthy adults and patients. For example, RR is a physiological state opposite to the state of stress response. Meditation is one effective way to induce RR and RR practice also includes a meditation component. We will discuss the effects of RR and meditation on gene expression. A cross-sectional study showed that 8 weeks of short-term RR practice induced genomic counter-stress changes and metabolism (Dusek et al., 2008). To test the long and short-term effects of the RR, a further study compared differences in gene expression changes after one RR session in expert meditators with years of experience and novices with only 8 weeks of practice. Results showed that experts and novices had different gene expression profiles at baseline, but experts had more consistent and pronounced gene expression changes after a RR session. Both groups presented upregulated gene changes related to energy metabolism, mitochondrial function, insulin secretion, and telomere maintenance, and downregulated genes changes associated with inflammatory responses and stress pathways (Bhasin et al., 2013). Studies also explored the clinical and genomic effects in patients with irritable bowel syndrome (IBS) and inflammatory bowel disease (IBD) following a 9 weeks of RR based mind—body intervention including meditation, yoga, and other techniques. IBS and IBD are chronic diseases of the digestive system which are exacerbated with stress. A pilot study showed that IBS and IBD patients showed greater quality of life, improved coping with pain, and significant reduction of symptoms and anxiety. Gene expression changes suggested that NF-κB is a key molecule in both IBS and IBD and that its regulation may contribute to reducing stress effects. However, given the uncontrolled design and small sample, future research should employ a RCT design with a larger sample to confirm these preliminary findings (Kuo et al., 2015).

A cross-sectional study in older adults showed that 8 weeks of MBSR reduced self-reported loneliness and proinflammatory gene expression patterns (Creswell et al., 2012). Another cross-sectional study explored the immediate effects of an intensive 8-h meditation retreat in experienced meditators on three sets of gene expression. Results showed that there were no differences in the tested gene expression between expert and naïve groups before meditation, but after meditation there was a significant silencing effect on several proinflammatory genes and histone deacetylase genes only in experienced meditators (Kaliman et al., 2014).

In a longitudinal study, investigators examined whether a 6-day residential meditation retreat had an impact on gene expression. The study had three groups: novice meditators and experienced meditators who attended about 8 h/day meditation and yoga practice, and an active control group that resided at the same location for the same amount of time, but did not participate in meditation. This control group was necessary because people participating in the retreat are not only meditating, but are also away from their daily workload, which, in theory, would lower their stress levels and may change gene expression. Results indicated that all psychological outcomes such as depression, stress, vitality, and mindfulness improved for all groups after the 6-day intervention and remained so at the 1-month follow-up. There was lower expression of genes related to stress response, wound healing, and injury in all three groups, most likely due to the relaxation effects common across groups. Experienced meditators showed lower expression of genes involved in protein synthesis, viral expression, and infectious cycle, whereas the novice meditators did not. All three groups also had a higher $A\beta42/A\beta40$ ratio (index of lower risk of dementia, depression, and mortality) after the retreat, but an unexpected finding was that experienced meditators had shorter telomeres, which is associated with aging, diabetes, cardiovascular disease, and some types of cancer (Epel et al., 2016).

A further cross-sectional study aimed to explore the epigenetic response to long-term mindfulness meditation. A genome-wide screen of DNA methylation in peripheral blood leukocytes was performed and results showed differentially methylated regions (DMRs) in long-term meditators compared to matched controls, such that about 70% of the mindfulness-related DMRs were hypomethylated in meditators and about 23% of DMRs clustered in telemetric chromosomal regions. Most importantly, almost 50% of the mindfulness-related DMRs involved genes linked to common human diseases, such as neurological and psychiatric disorders, cardiovascular diseases, and cancer. Further analysis revealed tumor necrosis factor and nuclear factor kappa-light-chain-enhancer of activated B cells (NF-κB) signaling are crucial regulators of the mindfulness-related genes. These results suggest that there is an epigenetic response to long-term meditation practice in blood leukocytes, involving genes linked to common human diseases. Further research is warranted to confirm these results using RCT design with a larger sample (García-Campayo et al., 2018).

Another review summarized diverse mind−body interventions (i.e., meditation, RR, yoga, tai chi, qigong, and breath regulation) on gene expression, and results suggested that these practices are associated with a downregulation of NF-κB-targeted genes via the nuclear factor kappa B pathway, which is the opposite to the effects of chronic stress on gene expression and may reduce the risk of inflammation-related diseases (Buric, Farias, Jong, Mee, & Brazil, 2017). Although these pilot studies have explored meditation influences on gene expression and identified candidate genes and biological pathways that may be sensitive to meditation, the methodological issues such as uncontrolled trials, cross-sectional design, and small sample size raise concerns. Future RCT studies with active controls are warranted to replicate and confirm these findings (Kaliman, 2018).

Reversibility of DNA methylation and/or histone modification

As we know, epigenetics involves heritable phenotype changes in gene activity and expression through DNA methylation and/or histone modification. The growing literature has shown how everyday life experiences and environment influence our gene expression. If DNA methylation plays a role in responding to experience in fully differentiated tissue, it must be biochemically reversible. That is, DNA should be either methylated or demethylated in response to environmental changes (Szyf et al., 2016; see also García-Campayo et al., 2018). Therefore DNA methylation is potentially preventable and reversible. Reversibility of DNA methylation is also critical for interventions such as meditation that may be relevant for resetting epigenetic programming. Although the mechanisms responsible for demethylation are still unclear, there is evidence that DNA methylation is potentially reversible even in mature and fully differentiated neurons (Szyf et al., 2016). These results have two important implications. First, DNA methylation could change in adult tissues and, therefore, adult tissue could be potentially responsive to environmental cues and readjustment of phenotypes. Second, it should be possible to intervene to reverse DNA methylation and alleviate adverse phenotypes (Szyf et al., 2016). In addition, DNA methylation markers might provide an objective tool for assessing effects of adverse experience on individual risks, as well as progress of an intervention. Despite this promising potential of translational epigenetics, many practical challenges must be addressed before behavioral epigenetics could become translational epigenetics (Szyf et al., 2016).

So far, the mechanisms of epigenetic changes following meditation remain mostly unknown, here we propose a putative framework or model——reversible epigenetics by meditation as shown in Fig. 4.1—with the hope that our proposal could motivate future research in the field of contemplative sciences and epigenetics (see also Black, Christodoulou, & Cole, 2019). Environment/Experience (E/E) such as stressor triggers CNS/ANS response (e.g., neuro-endocrinology, neuro-immunology), then induces cellular changes and gene expression. These accumulated effects could further

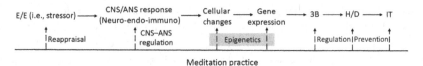

FIGURE 4.1   Reversible epigenetics by meditation framework.

affect brain, behavior and biology (3B), maintain health or lead to disorders (H/D) and induce intergenerational transmissions (IT). Meditation practice as a tailored experience has multiple influences on these above processes, for instance, meditation could affect stress response through reappraisal; meditation could modulate neuro-endocrinology and neuro-immunology via CNS/ ANS; meditation could have epigenetic effects on our molecular level; meditation also could change our behavior, brain and biology through self-regulation; moreover, meditation practice could prevent potential health problems and disorders and likely lead to reversible epigenetics. Central nervous system/autonomic nervous system: CNS/ANS

Intergenerational transmissions

Epigenetic changes occur over time in response to environment and experience such as stress, fear, mood disorders, addiction, and intervention. Animal studies indicate that certain conditional fear (e.g., electric shock) can be inherited and transmitted to the offspring. These epigenetic changes last up to two generations without reintroducing the shock (Szyf, 2014). This phenomenon is often called "intergenerational transmission." In general, intergenerational transmission refers to the transfer of traits from parents to offspring, including genetic and nongenetic influences. For example, epigenetic or behavioral changes in the offspring could be intergenerationally transmitted through prenatal effects and postnatal effects such as parent nutrition, stress, and other environmental factors (Ho, Sanders, Gotlib, & Hoeft, 2016). We will select stress and mood disorders such as depression as well as neural circuits as examples to explore intergenerational transmission and its reversibility.

Similar to epigenetic changes by conditional fear in animals, human data also indicate that offspring are affected by parental stress exposure or traumatic experiences. Physical, psychological, behavioral outcomes, and biological correlates such as neuroendocrine, epigenetic, and brain changes are observed in affected offspring (Bowers & Yehuda, 2016). For instance, a growing body of literature has demonstrated an intergenerational concordance between mother—child diurnal cortisol production. If mothers have a history of depression, intergenerational transmission of cortisol production could contribute to hypercortisolemia in children of depressed mothers, which has been shown to increase the risk for MDD (LeMoult, Chen, Foland-Ross, Burley, & Gotlib, 2015).

Parental stress-mediated effects in offspring could be explained by genetics or social learning theory. Parents' stress vulnerability stems from: (1) genetic risk factors such that offspring may inherit the genetic risks that have an impact on their own stress vulnerability; or (2) behavioral alterations by stress-related psychopathology that may affect the ability of parent or the childhood environment of offspring. Recent evidence suggests the potential epigenetic mechanisms——offspring of stress-exposed parents are at risk for adverse outcomes because of enduring epigenetic changes in parental biological systems and intergenerational transmission of the risk (Bowers & Yehuda, 2016). One of the adverse outcomes is the intergenerational transmission of risk for depression or anxiety. For example, research has shown that having a depressed mother is one of the strongest predictors for developing depression in adolescence. Moreover, having parents with mood disorders such as depression or anxiety increases the risk of these disorders in children and adolescents (Bowers & Yehuda, 2016; Colich et al., 2017; LeMoult et al., 2015).

Intergenerational transmission of neural circuits

At each moment, our brain responds to external environment and internal experiences and regulates our attention, emotion, cognition and social behavior through CNS and ANS. Certainly there is an intergenerational transmission of neural circuits including brain functional and structural changes in activation, white, or gray matter. Given the role of aberrant reward-processing in the onset and maintenance of depression, one study examined the association between mothers' and their daughters' neural response to the anticipation of reward and loss using functional magnetic resonance imaging (fMRI) in two groups——nondepressed mothers with a history of recurrent depression and their never-disordered daughters, and mothers without past or current depression and their never-disordered daughters. A significant association between mothers' and daughters' putamen (one area of reward system) response to the anticipation of loss was detected, regardless of the mother's depression history. Pubertal stage moderated the association between mother–daughter putamen concordance. These findings suggest a unique role of the putamen in the maternal transmission of reward learning and disturbances associated with depression (Colich et al., 2017). Another study compared three groups——currently depressed daughters of mothers with a history of MDD, matched never-depressed daughters of mothers with a history of MDD (high-risk), and matched healthy daughters——mothers control using task fMRI. Results showed that ventral striatum (part of reward system) activation was reduced for both currently depressed and high-risk girls compared to healthy controls. These results suggest major depression in mothers predicts reduced ventral striatum activation in offspring with and without depression (Sharp et al., 2014). Similar to brain functional changes, brain gray matter structure also demonstrates the same direction. A study included two groups: mothers with a history of depression (remitted mothers) and their never-depressed daughters (high-risk), and never-depressed mothers and their never-depressed daughters (healthy controls). Results showed that both remitted mothers and their high-risk daughters had thinner cortical gray matter compared to healthy controls. The extent of thickness anomalies in remitted mothers predicted analogous abnormalities in their daughters (Foland-Ross, Behzadian, LeMoult, & Gotlib, 2016). Is it possible to prevent or even reverse intergenerational transmission using meditation practice?

Prevention and reversibility of intergenerational transmission

Previous research has suggested disturbances in reward learning and processing are associated with a wide range of behavioral problems and mental disorders and their intergenerational transmission. The striatum plays a key role in the reward system and contributes to reward learning and processing, habit formation, and self-control processes, suggesting striatum may be a key brain region relevant for intergenerational transmission (Tang, 2017). Given that the activity and gray/white matter of striatum (e.g., putamen, caudate, and nucleus accumbens) often show reduction in patients with behavioral problems and mental disorders compared to healthy controls, it is possible to apply meditation to increase activity and gray/white matter of striatum and form new habits to treat these problems and disorders. The growing literature indicates that meditation has been successfully applied in the prevention and treatment of behavioral problems and mental disorders such as mood disorders, stress-related disorders, and addictions (Tang et al., 2015; Tang, Tang, & Gross, 2019; Tang, Tang, Rothbart et al., 2019). For example, in our RCT study, healthy college students without any training experience were randomly assigned to experimental (IBMT) or active control (relaxation training; RT) groups following five 30-minute of training. Compared to RT, brain imaging results showed increased activity and metabolism in striatum (e.g., putamen, caudate) and anterior cingulate cortex (ACC) after 5-session of IBMT. Behavioral and physiological data also demonstrated reduced stress hormone cortisol and negative moods and increased immune function, positive moods, and pleasant-reward experiences, suggesting that even brief periods of meditation could exert positive effects on health and well-being (Tang et al., 2009; Tang, Tang, & Gross, 2019; Tang, Tang, Rothbart et al., 2019).

To explore the underlying mechanisms, in another RCT study we measured the interaction between body (indexed by sympathetic−parasympathetic activity of autonomic nervous system, ANS) and brain (indexed by neural activity of central nervous system, CNS) following brief IBMT, and found that IBMT reduces stress hormones and improves immune function through downregulating the activity of major-stress axes in the body such as HPA axes, and strengthening brain self-control via the ACC and striatum (Tang, 2017). Following these results in the healthy population, we then extended the study to depression and anxiety patients using IBMT. Compared to the treatment as usual condition, a significant reduction of self-reported depression and anxiety scores and symptoms were found following 20 sessions of IBMT (30 min per session and 10 h in total). Self-control abilities measured by an attention network test and emotion regulation were significantly improved. Moreover, the abnormalities of brain activity and metabolism in striatum and ACC before IBMT were also changed to close to normal after 20 sessions of IBMT (Tang, 2017). These findings provide promising means to apply meditation in prevention and intervention of diverse behavioral problems and mental disorders and may reverse their intergenerational transmission. It should be noted that short and long-term meditation may have different impacts and lasting effects on intergenerational transmission given that short-term meditation often changes states, but long-term meditation can change traits (Tang, Holzel, & Posner, 2016). These studies seem to suggest that intergenerational transmission of traits is more stable than that of states because traits become relatively stable habits. Of note, other no-pharmacological approaches such as neurofeedback and cognitive training and exercise have also shown effects in reducing stress and symptoms and thus delaying, or even preventing, the onset of disorders.

In summary, generational effects have been seen in animal and human studies. The experiences of one generation can alter the behavior of the subsequent offspring. Such effects on cellular, physiological, and neural traits could result from lifespan experiences or environmental factors such as early life stress, depression and anxiety, addiction, and intervention. Therefore either intergenerational transmission is a simple transfer of the negative consequences or a way to enhance adaptive capacities in offspring. Overall, epigenetic information carried down across generations reduces developmental plasticity and flexibility and direct the development of offspring in a particular direction. This may help faster adaptation of offsprings to a new environment and increase the population size. However, offspring will be maladapted if the "anticipated" environment does not match the reality (Horsthemke, 2018; Lahiri et al., 2016).

In this chapter we discuss how genes affect and moderate behavior and meditation. We also discuss recent work on how short-term and long-term meditation experiences induce epigenetic changes in adults, which is promising for understanding how practice experience could have a lasting imprint on genes, illustrating the interplay of nature and nurture. Finally, we provide practice tips on meditation learning and practice based on translational and applied research findings. Although there are not many rigorous RCT studies related to genes, epigenetics, and meditation, we envision a promising future for clinical applications that use genetic information to guide meditation learning and practice, as well as developing or tailoring meditation to optimize its beneficial effects on human performance.

References

Albert, D., et al. (2015). Developmental mediation of genetic variation in response to the Fast Track prevention program. *Development and Psychopathology, 27*, 81–95.

Badran, B. W., Austelle, C. W., Smith, N. R., Glusman, C. E., Froeliger, B., Garland, E. L., … Short, B. (2017). A double-blind study exploring the use of transcranial direct current stimulation (tDCS) to potentially enhance mindfulness meditation (E-meditation). *Brain Stimulation, 10*(1), 152–154.

Bakermans-Kranenburg, M. J., Van IJzendoorn, M. H., Pijlman, F. T. A., Mesman, J., & Juffer, F. (2008). Experimental evidence for differential susceptibility: Dopamine D4 receptor polymorphism (DRD4 VNTR) moderates intervention effects on toddlers' externalizing behavior in a randomized controlled trial. *Developmental Psychology, 44*, 293–300.

Bhasin, M. K., Dusek, J. A., Chang, B. H., Joseph, M. G., Denninger, J. W., Fricchione, G. L., … Libermann, T. A. (2013). Relaxation response induces temporal transcriptome changes in energy metabolism, insulin secretion and inflammatory pathways. *PLoS ONE, 8*(5), e62817.

Bishop, J. R., Lee, A. M., Mills, L. J., Thuras, P. D., Eum, S., Clancy, D., … Lim, K. O. (2018). Methylation of FKBP5 and SLC6A4 in relation to treatment response to mindfulness based stress reduction for posttraumatic stress disorder. *Frontiers in Psychiatry, 9*, 418.

Black, D. S., Christodoulou, G., & Cole, S. (2019). Mindfulness meditation and gene expression: A hypothesis-generating framework. *Current Opinion in Psychology, 28*, 302–306.

Bowers, M. E., & Yehuda, R. (2016). Intergenerational transmission of stress in humans. *Neuropsychopharmacology, 41*, 232–244.

Brody, G. H., et al. (2014). Differential sensitivity to prevention programming: A dopaminergic polymorphism-enhanced prevention effect on protective parenting and adolescent substance use. *Health Psychology, 33*, 182–191.

Bronner, F., & Helmtrud, I. (Eds.), (2011). *Epigenetic aspects of chronic diseases*. Berlin: Springer.

Buric, I., Farias, M., Jong, J., Mee, C., & Brazil, I. A. (2017). What is the molecular signature of mind-body interventions? A systematic review of gene expression changes induced by meditation and related practices. *Frontiers in Immunology, 8*, 670.

Caspi, A., Sugden, K., Moffitt, T. E., Taylor, A., Craig, I. W., Harrington, H., … Poulton, R. (2003). Influence of life stress on depression: Moderation by a polymorphism in the 5-HTT gene. *Science, 301*(5631), 386–389.

Champagne, F. A., & Mashoodh, R. (2012). Genes in context: Gene-environment interplay and the origins of individual differences in behaviour. *Current Directions in Psychological Science, 18*(3), 127–131.

Cleveland, H. H., et al. (2015). The conditioning of intervention effects on early adolescent alcohol use by maternal involvement and dopamine receptor D4 (DRD4) and serotonin transporter linked polymorphic region (5-HTTLPR) genetic variants. *Development and Psychopathology, 27*, 51–67.

Coker, E. S., Gunier, R., Huen, K., Holland, N., & Eskenazi, B. (2018). DNA methylation and socioeconomic status in a Mexican-American birth cohort. *Clinical Epigenetics, 10*, 61.

Colich, N. L., Ho, T. C., Ellwood-Lowe, M. E., Foland-Ross, L. C., Sacchet, M. D., LeMoult, J. L., & Gotlib, I. H. (2017). Like mother like daughter: Putamen activation as a mechanism underlying intergenerational risk for depression. *Social, Cognitive, and Affective Neuroscience, 12*(9), 1480−1489.

Creswell, J. D., Irwin, M. R., Burklund, L. J., Lieberman, M. D., Arevalo, J. M., Ma, J., et al. (2012). Mindfulness-based stress reduction training reduces loneliness and proinflammatory gene expression in older adults: A small randomized controlled trial. *Brain, Behavior, and Immunity, 26*(7), 1095−1101.

Ding, X., Tang, Y. Y., Deng, Y., Tang, R., & Posner, M. I. (2015). Mood and personality predict improvement in creativity due to meditation training. *Learning and Individual Differences, 37*, 217−221.

Ding, X., Tang, Y. Y., Tang, R., & Posner, M. I. (2014). Improving creativity performance by short-term meditation. *Behavioral and Brain Functions, 10*, 9.

Dusek, J. A., Out, H. H., Wohlhueter, A. L., Bhasin, M., Zerbini, L. F., Joseph, M. G., & Libermann, T. A. (2008). Genomic counter-stress changes induced by the relaxation response. *PLoS One, 3*(7), e2576.

Epel, E. S., Puterman, E., Lin, J., Blackburn, E. H., Lum, P. Y., Beckmann, N. D., et al. (2016). Meditation and vacation effects have an impact on disease-associated molecular phenotypes. *Translational Psychiatry, 6*(8), e880.

Foland-Ross, L. C., Behzadian, N., LeMoult, J., & Gotlib, I. H. (2016). Concordant patterns of brain structure in mothers with recurrent depression and their never-depressed daughters. *Developmental Neuroscience, 38*(2), 115−123.

Francis, R. C. (2012). *Epigenetics: How environment shapes our genes.* W. W. Norton & Company.

García-Campayo, J., Puebla-Guedea, M., Labarga, A., Urdanoz, A., Roldan, M., Martinez de Morentin, X., . . . Mendioroz, M. (2018). Epigenetic response to mindfulness in peripheral blood leukocytes involves genes linked to common human diseases. *Mindfulness, 9*(4), 1146−1159.

Ho, T. C., Sanders, S. J., Gotlib, I. H., & Hoeft, F. (2016). Intergenerational neuroimaging of human brain circuitry. *Trends in Neuroscience, 39*(10), 644−648.

Horsthemke, B. (2018). A critical view on transgenerational epigenetic inheritance in humans. *Nature Communications, 9*, 2973.

Isbell, E., Stevens, C., Pakulak, E., Hampton Wray, A., Bell, T. A., & Neville, H. J. (2017). Neuroplasticity of selective attention: Research foundations and preliminary evidence for a gene by intervention interaction. *Proceedings of the National Academy of Sciences of the United States of America, 114*(35), 9247−9254.

Kaliman, P. (2018). Epigenetics and meditation. *Current Opinion in Psychology, 28*, 76−80.

Kaliman, P., Alvarez-Lopez, M. J., Cosín-Tomás, M., Rosenkranz, M. A., Lutz, A., & Davidson, R. J. (2014). Rapid changes in histone deacetylases and inflammatory gene expression in expert meditators. *Psychoneuroendocrinology, 40*, 96−107.

Kitayama, S., King, A., Yoon, C., Tompson, S., Huff, S., & Liberzon, I. (2014). The Dopamine D4 Receptor Gene (DRD4) moderates cultural difference in independent versus interdependent social orientation. *Psychological Science, 25*(6), 1169−1177.

Kuo, B., Bhasin, M., Jacquart, J., Scult, M. A., Slipp, L., Riklin, E. I., . . . Denninger, J. W. (2015). Genomic and clinical effects associated with a relaxation response mind-body intervention in patients with irritable bowel syndrome and inflammatory bowel disease. *PLoS ONE, 10*(4), e0123861.

Kuzelova, H., Ptacek, R., & Macek, M. (2010). The serotonin transporter gene (5-HTT) variant and psychiatric disorders: Review of current literature. *Neuro Endocrinology Letters, 31*(1), 4−10.

Lahiri, D. K., Maloney, B., Bayon, B. L., Chopra, N., White, F. A., Greig, N. H., & Nurnberger, J. I. (2016). Transgenerational latent early-life associated regulation unites environment and genetics across generations. *Epigenomics*, *8*(3), 373−387.

Lam, L. L., Emberly, E., Fraser, H. B., Neumann, S. M., Chen, E., Miller, G. E., & Kobor, M. S. (2012). Factors underlying variable DNA methylation in a human community cohort. *Proceedings of the National Academy of Sciences of the United States of America* (Suppl 2), 17253−17260.

LeMoult, J., Chen, M. C., Foland-Ross, L. C., Burley, H. W., & Gotlib, I. H. (2015). Concordance of mother-daughter diurnal cortisol production: Understanding the intergenerational transmission of risk for depression. *Biological Psychology*, *108*, 98−104.

Miller, G. (2010). Epigenetics. The seductive allure of behavioral epigenetics. *Science*, *329* (5987), 24−27.

Moore, D. S. (2015). *The developing genome: An introduction to behavioral epigenetics* (1st ed.). Oxford University Press.

Plak, R. D., Kegel, C. A., & Bus, A. G. (2015). Genetic differential susceptibility in literacy-delayed children: A randomized controlled trial on emergent literacy in kindergarten. *Development and Psychopathology*, *27*(1), 69−79.

Puig, M. V., & Gulledge, A. T. (2011). Serotonin and prefrontal cortex function: Neurons, networks, and circuits. *Molecular Neurobiology*, *44*, 449−464.

Renthal, W., & Nestler, E. J. (2008). Epigenetic mechanisms in drug addiction. *Trends in Molecular Medicine*, *14*(8), 341−350.

Saghazadeh, A., Esfahani, S. A., & Rezaei, N. (2016). Genetic polymorphisms and the adequacy of brain stimulation: State of the art. *Expert Review of Neurotherapeutics*, *16*(9), 1043−1054.

Sharp, C., Kim, S., Herman, L., Pane, H., Reuter, T., & Strathearn, L. (2014). Major depression in mothers predicts reduced ventral striatum activation in adolescent female offspring with and without depression. *Journal of Abnormal Psychology*, *123*(2), 298−309.

Sheese, B. E., Rothbart, M. K., Voelker, P. M., & Posner, M. I. (2012). The dopamine receptor D4 gene 7-repeat allele interacts with parenting quality to predict effortful control in four-year-old children. *Child Development Research*, *2012*, 863242.

Sheese, B. E., Voelker, P. M., Rothbart, M. K., & Posner, M. I. (2007). Parenting quality interacts with genetic variation in dopamine receptor D4 to influence temperament in early childhood. *Development and Psychopathology*, *19*(4), 1039−1046.

Smith, H. J., Kryski, K. R., Sheikh, H. I., Singh, S. M., & Hayden, E. P. (2013). The role of parenting and dopamine D4 receptor gene polymorphisms in children's inhibitory control. *Developmental Science*, *16*(4), 515−530.

Smith, H. J., Sheikh, H. I., Dyson, M. W., Olino, T. M., Laptook, R. S., Durbin, C. E., ... Klein, D. N. (2012). Parenting and child DRD4 genotype interact to predict children's early emerging effortful control. *Child Development*, *83*(6), 1932−1944.

Sumirtanurdin, R., Thalib, A. Y., Cantona, K., & Abdulah, R. (2019). Effect of genetic polymorphisms on Alzheimer's disease treatment outcomes: An update. *Clinical Interventions in Aging*, *14*, 631−642.

Szyf, M. (2014). Lamarck revisited: Epigenetic inheritance of ancestral odor fear conditioning. *Nature Neuroscience*, *17*(1), 2−4.

Szyf, M., Tang, Y. Y., Hill, K. G., & Musci, R. (2016). The dynamic epigenome and its implications for behavioral interventions: a role for epigenetics to inform disorder prevention and health promotion. *Translational Behavioral Medicine*, *6*(1), 55−62.

Tang, Y. Y. (2017). *The neuroscience of mindfulness meditation: how the body and mind work together to change our behavior.* Cham: Springer Nature.

Tang, Y. Y., & Posner, M. I. (2009). Attention Training and Attention State training. *Trends in Cognitive Sciences, 13*(5), 222–227.

Tang, Y. Y., Ma, Y., Fan, Y., Feng, H., Wang, J., Feng, S., & Fan, M. (2009). Central and autonomic nervous system interaction is altered by short term meditation. *Proceedings of the National Academy of Sciences, USA, 106*(22), 8865–8870.

Tang, Y. Y., Posner, M. I., Rothbart, M. K., & Volkow, N. D. (2015). Circuitry of self-control and its role in reducing addiction. *Trends in Cognitive Sciences, 19*(8), 439–444.

Tang, Y. Y., Holzel, B. K., & Posner, M. I. (2016). Traits and states in mindfulness meditation. *Nature Reviews Neuroscience, 17*(1), 59.

Tang, Y. Y., Tang, R., & Gross, J. J. (2019). Promoting psychological well-being through an evidence-based mindfulness training program. *Frontiers in Human Neuroscience, 13*, 237.

Tang, Y. Y., Tang, R., Rothbart, M. K., & Posner, M. I. (2019). Frontal theta activity and white matter plasticity following mindfulness meditation. *Current Opinions in Psychology, 28*, 294–297.

Tolahunase, M. R., Sagar, R., Faiq, M., & Dada, R. (2018). Yoga- and meditation-based lifestyle intervention increases neuroplasticity and reduces severity of major depressive disorder: A randomized controlled trial. *Restorative Neurology and Neuroscience, 36*(3), 423–442.

Tompson, S. H., Huff, S. T., Yoon, C., King, A., Liberzon, I., & Kitayama, S. (2018). The dopamine D4 receptor gene (DRD4) modulates cultural variation in emotional experience. *Culture and Brain., 6*(2), 118–129.

Woo, H. I., Lim, S. W., Myung, W., Kim, D. K., & Lee, S. Y. (2018). Differentially expressed genes related to major depressive disorder and antidepressant response: Genome-wide gene expression analysis. *Experimental & Molecular Medicine, 50*(8), 92.

Zhang, T. Y., & Meaney, M. J. (2010). Epigenetics and the environmental regulation of the genome and its function. *Annual Review of Psychology, 61*, 439–466.

Chapter 5

Sympathetic and parasympathetic systems in meditation

Our central nervous system (CNS) includes the brain and spinal cord as well as the peripheral nervous system, which is subdivided into the somatic and autonomic systems. The autonomic nervous system (ANS) is further divided into sympathetic and parasympathetic systems that work together to maintain a state of homeostasis in the body. The ANS can be regarded as a control system that acts largely unconsciously and autonomically to regulate fundamental bodily functions such as heart rate (HR), respiratory rate, and digestion. Within the brain, the hypothalamus, an integrator for autonomic functions, regulates the ANS through receiving regulatory input from the somatosensory and limbic systems. Within ANS the main function of the sympathetic system is to mobilize and innervate the body's response under stressful situations, engaging organs and body parts such as the eyes, lungs, kidneys, gastrointestinal tract, and heart. In contrast, the parasympathetic system is responsible for the "rest and digest" phase of the body and involves functions that do not require an immediate reaction. These two systems often act in an opposing manner. For example, the parasympathetic system decreases HR and reduces blood pressure, whereas the sympathetic system increases HR and increases blood pressure. However, traditional views that regard the parasympathetic system as inhibitory and the sympathetic system as excitatory have proven to be too simplistic because there are many exceptions in these two systems. The best characterization may be that the sympathetic system is a quick-response mobilizing system and the parasympathetic system is a more slowly activated dampening system (Pocock, 2006).

The ANS, which includes sympathetic and parasympathetic systems, regulates bodily functions such as HR, respiratory rate, and skin conductance response (SCR) and these indexes are often used to measure the sympathetic and parasympathetic activities of the ANS. Heart rate variability (HRV) is a physiological phenomenon of variation in the time interval between heartbeats (variation in the beat-to-beat interval). The frequency spectrum of HRV includes high frequency power (HF) with a frequency activity in the

The Neuroscience of Meditation. DOI: https://doi.org/10.1016/B978-0-12-818266-6.00006-X

range of 0.15—0.40 Hz and low frequency power (LF) with a frequency activity in the range of 0.04—0.15 Hz. Increased sympathetic activity or decreased parasympathetic activity often results in reduced HRV. HF HRV activity has been linked to parasympathetic activity. For example, HF HRV is related to a vagally mediated modulation of HR such that HR increases during inhalation and decreases during exhalation, suggesting HF HRV as a sensitive biomarker of parasympathetic activity. Although LF HRV was thought to reflect sympathetic activity, it is now widely accepted that it reflects a mixture of sympathetic and parasympathetic activities (Billman, 2013). Similarly, some researchers consider LF/HF ratio as an indicator of sympathetic to parasympathetic autonomic balance, but this view remains controversial (Quintana & Heathers, 2014). SCR (or galvanic skin response) is an indication of psychological or physiological arousal. When the sympathetic system is highly aroused, then sweat-gland activity increases, which, in turn, increases SCR. Reversely, lower SCR reflects greater parasympathetic activity. Therefore in our series of randomized controlled trials (RCTs) of meditation-based interventions, we mainly focused on HF HRV and SCR as indexes for assessing sympathetic and parasympathetic activities, although LF HRV and LF/HF ratios were also analyzed (Tang, 2017; Tang et al., 2009).

The effect of meditation on sympathetic and parasympathetic systems

Meditation practice emphasizes mental processes but the role of the body (physiology) has often been overlooked in the scientific research of meditation practices. However, body and mind can never be separated in meditation practices or, for that matter, in our daily lives. In addition to brain changes, meditation practice is also accompanied by ANS changes such as greater parasympathetic activity indexed by a lower HR, chest respiratory rate, and SCR. Furthermore, there are greater abdominal breath amplitude (BA) and HF HRV. An earlier study on meditation showed that mantra-based meditation induced physiological changes including lower oxygen consumption, HR, and skin resistance and more alpha waves as measured by electroencephalogram (EEG) (Wallace, 1970). Later studies have used these indexes of ANS activity, including HR, HRV, SCR, respiratory amplitude/rate, and EEG frequencies as biomarkers for monitoring meditative states (Tang, 2017; Tang et al., 2010).

As shown in a recent article on mindfulness of the body, one of the core techniques of mindfulness meditation is paying attention to bodily sensations and feelings (body-focused attention) (Tang,Tang, & Gross, 2019), for example, body scan, which is a meditative practice involving "moving a focused spotlight of attention from one part of the body to another." However, mindfulness practice that involves a somatic focus is still processed in the mind, but may amplify and strengthen the connection between bodily sensations, feelings, and mental processes (Tang & Tang, 2015; Tang et al., 2019). Interoception is the ability to notice and become aware of physiological signals originating inside the body, such as heartbeat and the sense of hunger or pain. It has often been reported that attending to present-moment body sensations in meditation can result in an enhanced awareness of bodily states and greater perceptual clarity of subtle interoception. Studies have also found that meditators showed greater coherence between objective physiological indexes and their subjective emotional experiences and sensitivity of body regions (Tang, Holzel, & Posner, 2015; Tang et al., 2019). Using functional neuroimaging, subjects were assessed while they judged the timing of their own heartbeats, researchers found that there were enhanced activity in insula, somatomotor, and cingulate cortices, while more sympathetic activity accompanied this task (Critchley, 2004). Although studies did not find evidence that meditators had superior performance than nonmeditators using the heartbeat detection task—a standard measure of interoceptive awareness, neural activity in the right anterior insular/opercular cortex predicted the accuracy of participants in the task (Tang et al., 2015). Another study suggested that the right insula seems to support the sympathetic activity during the heartbeat detection task (Lutz, Greischar, Perlman, & Davidson, 2009). Our study on integrative body−mind training (IBMT), a form of mindfulness meditation, showed that the IBMT group exhibited more left insula activity compared to the relaxation group, consistent with previous findings that the left insula is predominantly responsible for parasympathetic effects (Tang et al., 2010). Taken together these findings suggest that meditation can engage both sympathetic and parasympathetic activities, with the left insula related to parasympathetic activity and the right insula related to sympathetic activity.

Based on previous research we have proposed that mindfulness states can be achieved in two ways: through mental processes (e.g., mindfulness) and through bodily processes (e.g., bodifulness). Bodifulness refers to the gentle and holistic adjustment and exercise of body posture with a full awareness in order to achieve a presence, balance, and integration in our bodies (Tang & Tang, 2015; Tang et al., 2019). For instance, in eastern traditions, practices like IBMT, TCM-based methods (e.g., tai chi, qigong, yoga) emphasize body−mind balance and interaction to facilitate meditative states, thereby achieving positive outcomes (Tang, 2017; Tang & Tang, 2015). Mindfulness involves an explicit process (e.g., counting your breath) through CNS (brain/mind), but bodifulness mainly involves an implicit process (e.g., visceral or interoceptive awareness) regulated by ANS. Autonomic control needs less effort and is supported by the anterior cingulate cortex (ACC) and adjacent medial prefrontal cortex. In a review article we divided meditation into three different stages of meditation practice, namely early, middle (intermediate) and advanced, that involve different amounts of effort (Tang, Rothbart, &

Posner, 2012). The early stages of meditation require conscious cognitive control with effort and is often supported by the lateral prefrontal and parietal cortex. The more active lateral prefrontal and parietal cortex may reflect the higher level of mental effort when participants struggle to obtain the meditation state in the early stages. In this situation the sympathetic system likely plays a key role in supporting the early stages of meditation. However, in the advanced stages when practitioners become more skilled in meditation, prefrontal-parietal activity is often reduced or eliminated, but the activities of some brain regions such as ACC, striatum, and insula remain stable, suggesting the dominance of the parasympathetic system that requires less effort (Tang et al., 2015). Moreover, when the mental process of meditation through cognitive control becomes automatic and could be internalized through the body (via autonomic control), a state of bodifulness is formed. Cognitive and autonomic control are important components of self-control, which facilitate meditative states and underlie behavior and habit formation (Tang, 2017; Tang & Tang, 2015; Tang et al., 2019). Whether effort has a key role in sympathetic and parasympathetic systems during or following meditation needs further investigation.

Because meditation, such as IBMT, changes the state of the body through autonomic control, we hypothesize that ANS activity, especially parasympathetic activity, will increase during, and following, meditation. For example, in a randomized study including five sessions of experimental (IBMT) or control (relaxation) groups, we measured the HRV, SCR, respiratory amplitude, and rate——indexes of sympathetic and parasympathetic activity. During and after training, the IBMT group showed significantly better physiological reactions such as lower HR and more abdomen respiratory amplitude when compared with the relaxation control group. We also found that compared to relaxation training, IBMT significantly improved HF HRV and reduced SCR, suggesting better parasympathetic regulation. Of note is that IBMT does not intentionally exercise or change breathing patterns from chest to abdomen, instead practitioners naturally move toward the direction of a more abdomen respiratory pattern. Although our observations suggest that IBMT may initiate our innate autonomic system to induce these positive changes, the underlying mechanisms warrant further investigation (Tang, 2017; Tang et al., 2010). In the same vein, a comprehensive review of mind–body practice (e.g., yoga) showed the same effects——increased vagal tone (more HF HRV) that facilitates autonomic balance (Tyagi & Cohen, 2016). Taken together, meditation practices tend to increase parasympathetic activity as indexed by more HF HRV (increased vagal tone), abdomen respiratory amplitude, and lower SCR (Gerritsen & Band, 2018). It should be noted that if one devotes more effort and control into meditation practices, then the practices would likely be similar to performing a task with a high workload. Therefore it is possible that such practice may then induce more sympathetic activity. One study examined effort and SCR in a patient without a feeling of mental effort and suggested that SCR is a sensitive biomarker of effort (Naccache et al., 2005). This result indicates that we could use SCR to monitor and evaluate the amount of mental effort involved in meditation and assess the stage of meditators. Nevertheless, research examining the coordination of parasympathetic and sympathetic interaction is limited and data seem to suggest respiration patterns may play a key role in the regulation of both parasympathetic and sympathetic activities, which is in line with our IBMT findings showing the relationship between abdomen respiration and the parasympathetic system (Tang, 2017).

Brain waves and heart rate variability in sympathetic and parasympathetic systems

Given that increased sympathetic activity or decreased parasympathetic activity often results in reduced HRV, studies have investigated the relationship between HRV and sympathetic and parasympathetic activities. For instance, higher resting HRV is related to more adaptive and functional top-down and/or bottom-up cognitive modulation of emotional stimuli, which may facilitate effective emotion regulation. In contrast, lower resting HRV is associated with hypervigilant and maladaptive cognitive responses to emotional stimuli, which may be destructive to emotion regulation. These results suggest that increased parasympathetic activity may lead to better emotion regulation (Park & Thayer, 2014), a phenomenon that is often observed in many meditation findings, which could be the underlying mechanisms supporting enhanced emotion regulation following meditation (Tang et al., 2015).

What is the relationship between brain waves and sympathetic and parasympathetic activities? We will take sleep stages in newborns and adults as an example. Sleep stages include active sleep and quiet sleep, which are differentiated by EEG patterns, eye movements, respiratory patterns, and HRV. The results showed that in newborns all LF, HF, and the LF/HF HRV ratio during active sleep were significantly higher than those during quiet sleep, suggesting a special HRV pattern during this early development period. In adults the HF values during rapid eye movement (REM) sleep (active sleep) were significantly lower than those during non-REM sleep (quiet sleep) and the LF/HF ratio during active sleep was significantly higher than that during quiet sleep. REM sleep has also been found to have similar brain waves as those during wakefulness, such as beta $(12-38 \text{ Hz})$ with fast frequency and lower amplitude. Non-REM sleep has brain waves in the range of alpha $8-12$ hz stage 1, $3-8$ stage 2, delta stage 3, with slower frequencies and higher amplitudes corresponding to different sleep stages $1-3$. These results indicate that the autonomic patterns of brain and sympathetic and parasympathetic activities in active and quiet sleep of newborns are very different from those in REM and non-REM sleep of adults and may develop into autonomic patterns in adults. It is important to point out that in newborns, sympathetic and parasympathetic systems work together to support the rapid pace of development. In addition, in some preterm infants, the delayed development of the autonomic system can be determined by classifying the sympathetic and parasympathetic patterns of sleep stages (Takatani et al., 2018).

In our RCT studies we measured both brain activity and sympathetic and parasympathetic activities using multiple measures such as EEG, HRV, and SCR before, during and after meditation (Tang, 2017; Tang et al., 2009, 2015). We randomized college students into meditation and relaxation training groups and found no differences in Heart rate (HR), respiratory amplitude and rate, SCR, EEG power, and brain activity between the two groups at baseline. To reveal the dynamic changes during each training session, we divided each 30 min training session into periods 1, 2, and 3 (beginning, deep, and ending periods of 10 min each). After five sessions of training, the main effect of training session was significant for SCR in the meditation group compared to the relaxation group. The group x session interaction was significant. Further, in comparison with relaxation, SCR was significantly lower during practice periods 2 and 3, as well as after training completion when participants returned back to baseline, indicating that the SCR was related to the length of training (see Fig. 5.1B). Similar results were found for HR, abdomen respiratory amplitude, and chest respiratory rate. In particular, after five sessions of training, the main effect of the training session was significantly lower for HR, greater for abdomen respiratory amplitude, and lower for chest respiratory rate in the meditation group compared to the relaxation group. The group x session interaction was significant for HR, abdomen respiratory amplitude, and chest respiratory rate, respectively. We also performed an analysis of HRV. The main effect of the training session was significant for the percentage of change in the high-frequency (HF) HRV in the meditation group compared to the relaxation group, and the group x session interaction was also significant. In comparison with the relaxation group, HF HRV in the meditation group was significantly improved during practice period 2 and marginally significant during practice period 3, respectively, indicating that HF HRV was more related to the depth of meditation state (see Fig. 5.1A). The results of lower SCR, increased abdomen respiratory amplitude, decreased chest respiratory rate, and more HF HRV demonstrated better ANS regulation, especially more parasympathetic activity during and following meditation in comparison with relaxation training (Tang et al., 2009).

To explore the underlying brain mechanisms during short-term meditation in the same groups, we recorded brain activity using EEG. Before training, none of the scalp electrodes of the EEG showed differences between the two groups. However, in the meditation group, a group x session effect for frontal midline electrodes Fz, FCz, and Cz were detected, respectively. After training, we also found significant increases in the meditation group (but not

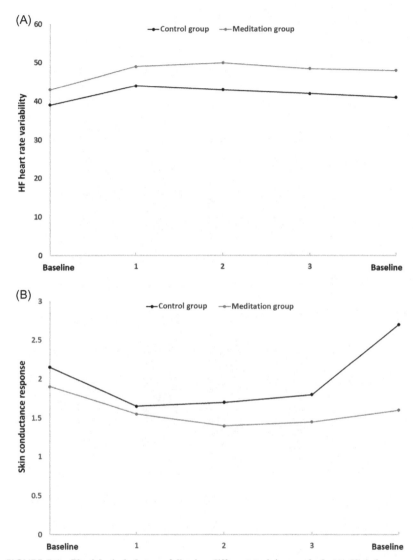

FIGURE 5.1   Physiological changes following different training methods (A) High-frequency HRV following meditation group (IBMT) and control group (relaxation training). (B) Skin conductance response (SCR) following meditation group (IBMT) and control group (relaxation training). (C) SCR and nuHF HRV following mindfulness meditation (IBMT) and physical exercise (PE). (D) The resting heart rate (HR) change and percentage of change of chest breath amplitude (BA) following IBMT and PE.

the relaxation group) in EEG power in the theta frequency band (3−8 Hz) for frontal midline electrodes Fz, FCz, and Cz, which are often related to the source of the ACC (Tang et al., 2015). These findings indicated that merely

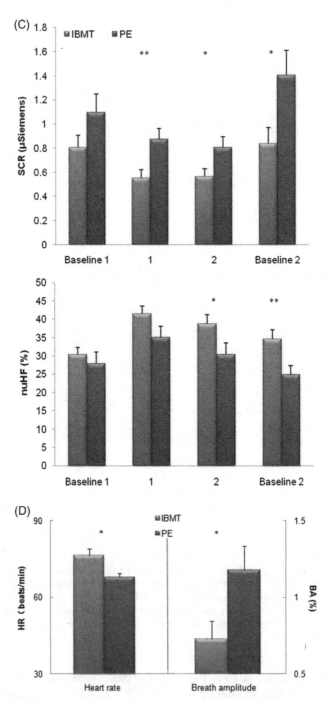

FIGURE 5.1   (Continued)

five sessions of meditation can induce positive changes in the brain and sympathetic and parasympathetic systems. Furthermore, these findings suggest that the parasympathetic activity and midline frontal theta power may be biomarkers relevant for meditative states and outcomes.

Body–mind interaction during meditation

A series of studies have indicated that meditation regulates ANS function and induces physiological changes such as oxygen consumption, HR, respiratory amplitude and rate, SCR, HRV, and other indexes. These findings suggest the dominance of parasympathetic activity following meditation practice. Moreover, the awareness of bodily sensations and the practice of meditation techniques focusing on full awareness and presence of the body (bodifulness) also facilitate the mindfulness aspect of meditation practices. This notion is in line with the embodiment literature that body posture and state change mental processes such as emotional processing, the retrieval of autobiographical memories, and cortisol concentrations (Dijkstra, Kaschak, & Zwaan, 2007; Hennig et al., 2000; Huang, Galinsky, Gruenfeld, & Guillory, 2011; Niedenthal, 2007; Tang et al., 2019). In other words, our body postures change our chemistry, our bodies change our minds, and our minds change our behavior (Tang, 2017; Tang et al., 2015).

Does meditation change neurophysiology? A systematic review of EEG studies of mindfulness meditation examined EEG power outcomes in each bandwidth, in particular the power differences between meditation state and control state. It also examined outcomes relating to hemispheric asymmetry and event-related potentials (ERP). The results suggested that the meditation state was most commonly related to enhanced alpha and theta power when compared to an eyes-closed resting state. However, there were no consistent patterns observed in beta, delta, and gamma bandwidths. This copresence of elevated alpha and theta power may signify a state of relaxed alertness and calmness which together leads to health and well-being (Lomas, Ivtzan, & Fu, 2015).

To further study the mechanisms of body–mind interaction, we applied brain imaging and physiology measures after five sessions of IBMT (mindfulness meditation) and relaxation training in two randomized studies. We first found stronger subgenual/ventral ACC activity in the IBMT group (Tang et al., 2010). Given that this brain area has also been linked to ANS activity (Bush, Luu, & Posner, 2000; Critchley, 2004; Posner, Sheese, Rothbart, & Tang, 2007), we then measured the HRV and SCR, two indexes of sympathetic and parasympathetic activities. During and after training the IBMT group showed significantly better physiological reactions in HR and respiratory amplitude and rate compared to the relaxation control. We also found that compared to relaxation training, IBMT significantly improved HF HRV and reduced SCR, suggesting better parasympathetic regulation (see Fig. 5.1A and B), while EEG power showed greater ACC theta brain activity. Frontal midline ACC theta power correlated with HF HRV, suggesting control by the ACC over parasympathetic activity. These results indicate that after five sessions of training, the meditation group showed better regulation of the ANS through a ventral midline brain system than the relaxation group. This changed brain state and activity probably reflect training in the coordination of body and mind through meditation, as these changes were not observed in the control group. Therefore ACC and ANS may serve as mediating brain mechanisms underlying meditation-related improvements in attention control and emotion regulation as well as other behavior (Tang, 2017; Tang et al.,2009).

From a practice perspective, meditation relying solely on mind control (without body engagement) often leads to a "dry" practice experience, that is, practitioners put lots of effort into control and usually find it difficult to achieve meditative states. As a result, this process is usually associated with mental fatigue, negative emotion, and difficulty in learning to meditate. This is consistent with previous findings that brain and mind work together to support meditation states (Kerr, Sacchet, Lazar, Moore, & Jones, 2013; Tang et al., 2007, 2015, 2019).

Body−mind interaction (or body−brain interaction) occurs following other meditation methods given that the parasympathetic system (e.g., breathing amplitude and rate, SCR, HRV) also engages the brain during meditation. A study explored the brain−heart interactions in a group of monks with many years of experience in traditional Tibetan Buddhist meditation using EEG and ECG (electrocardiogram, a recording of the electrical activity of the heart). The hypothesis was that meditation should reflect changes in the neural representations of visceral activity, such as cardiac behavior, and that meditation should also induce the integration of neural and visceral systems and change the spontaneous whole-brain spatiotemporal dynamics. Compared to the control group, the monk group showed different transient modulations of the neural response to heartbeats in the default mode network (DMN), along with large-scale network reconfigurations in the theta and gamma bands of EEG activity following meditation. The temporal-frontal network connectivity in the EEG theta band was negatively correlated with the duration of meditation experience, and gamma oscillations accompanied theta oscillations during meditation. Overall, these data support the hypothesis. However, one puzzle in the study is the gamma−theta relationship during meditation, since the gamma band was very fast whereas the theta band was slow. It remains unclear how these two bands work together to induce brain and heart changes (Jiang et al., 2019). The DMN has been consistently associated with self-related processing and also with autonomic regulation such as HR and HRV. Some prominent works have demonstrated that these two functions are coupled in the DMN since selfhood is grounded in the neural monitoring of internal organs

(Babo-Rebelo, Richter, & Tallon-Baudry, 2016). Another study examined the interaction between state (compassion vs neutral) and group (novice vs expert meditator) and their effect on the relation between HR and brain activity during the presentation of emotional sounds in each state. Results showed stronger positive coupling in the dorsal ACC between HR and brain activity during compassion meditation states compared to the neutral state (Lutz et al., 2009). Similarly, our study revealed a positive correlation between frontal midline theta ACC activity and HF HRV after five sessions of IBMT (Tang et al., 2010). Therefore the ACC appears to play a key role in the brain—heart interactions and network dynamics and seems to be critical for meditation practice and meditative states (Tang et al., 2015).

Anterior cingulate cortex As a bridge for autonomic nervous system and central nervous system interaction

Self-control is the ability to control one's emotion, cognition, and behavior in accordance with internal or external goals. The ACC is part of a network implicated in monitoring, controlling, and resolving conflict among competing response tendencies. During infancy and childhood, the ACC has been shown to alter its connectivity with other brain areas. These changes are related to the increasing ability of children to regulate their emotions and behavior (Posner et al., 2007). Moreover, the ACC is active in many studies of brain imaging, especially during executive attention, cognitive control, and self-regulation (Posner et al., 2007). We have proposed that a key role of the ACC is supporting self-control during meditation (Tang et al., 2015). The sensitivity of the ACC to reward and pain, and evidence for ACC coupling (functional connectivity) with areas related to cognitive and emotional processing during task performance, also supports this hypothesis. Likewise, the correlation between the efficiency of the executive attention network and the ratings of parents on the self-regulatory ability of their child also confirm this idea (Posner & Rothbart, 2006). In addition, deficits in ACC activity have been associated with many disorders related to self-control deficits, such as attention deficit/hyperactivity disorder, depression, dementia, addiction, and schizophrenia (Posner et al., 2007). For instance, in addiction literature, hypoactivation of the ACC has been found to be critical for symptoms of craving. In tobacco use, a circuit involving the ACC and striatum has been shown to have lower than normal connectivity (Tang et al., 2010, 2013). Given that ACC deficits are associated with many disorders, the ability to strengthen ACC activity and connectivity through training could provide a means for improving self-control and may serve as a possible therapy or prevention tool. In meditation research, ACC functional changes and structural changes (e.g., white-matter connectivity) have been shown. For example, five sessions of IBMT induce greater ACC activity and 10 sessions of IBMT changes brain white-matter structure. Moreover, white-matter changes surrounding the ACC are correlated with emotion regulation (Tang et al., 2012, 2015). Therefore these studies provide evidence that meditation can induce increased brain activity and strengthen the connectivity of the ACC, a region critical for self-control, suggesting that meditation could potentially be used for treating or preventing problems and disorders associated with deficits of self-control such as addiction, mood disorders, ADHD, and PTSD (Tang, 2017; Tang et al., 2010, 2013).

Why is the anterior cingulate cortex involved in so many functions?

The ACC is the frontal part of the cingulate cortex which looks like a "collar" surrounding the frontal part of the corpus callosum and connects the upper cerebral cortex and lower limbic areas. Anatomically, the ACC can be divided into dorsal and ventral parts, which are responsible for both cognitive and emotional functions, respectively (Bush et al., 2000). The dorsal ACC is responsible for processing top-down and bottom-up information and is connected to the prefrontal cortex and parietal cortex. In contrast, the ventral ACC is linked to the nucleus accumbens, hypothalamus, hippocampus, anterior insula, and amygdala that serves a role in assessing the salience of emotion and motivational information. Traditionally, rostral ACC is implicated in autonomic control. Human and animal research has demonstrated that ACC stimulation evokes a range of autonomic responses, and ACC appears to play a role in a wide variety of autonomic functions, such as regulating blood pressure, HR, and homeostasis (Critchley, 2004; Gizowski & Bourque, 2018). ACC is also involved in higher-level functions, such as attention allocation, cognitive control, emotion, reward anticipation, and decision making (Posner et al., 2007; Shackman et al., 2011). Most importantly, a large body of evidence has suggested that ACC is a crucial brain region that bridges the cerebral cortex and limbic system and can regulate ANS efficiently (Critchley, 2004; Posner et al., 2007; Shackman et al., 2011; Tang et al., 2010). In our series of IBMT studies we found that integrating bodifulness and mindfulness into mindfulness meditation could significantly improve the functions of ANS and CNS and change brain plasticity and physiology (Tang et al., 2019). For example, 10−20 sessions of IBMT (30 min per session) within 2−4 weeks induces brain structural changes in white matter surrounding the ACC. In particular, after 20-session (4-week) training with IBMT, improved white matter involves myelination and axonal changes. However, 10-session (2-week) IBMT training only induced axonal changes. These results demonstrate the time-course of white-matter neuroplasticity in short-term meditation (Tang et al., 2012). Furthermore, this dynamic pattern of white-matter change involving the ACC is somewhat similar to the order of white-matter changes found during development. Future studies should provide a means for a better understanding of the developmental process involving white matter and develop interventions to improve attention, emotion, and behavior or prevent related problems and disorders (Tang et al., 2012).

It should be noted that 10 to 20 sessions of IBMT induced dorsal and ventral ACC structural changes and changed white-matter tracts connecting the ACC to the striatum and other brain areas supporting self-control abilities. However, five sessions of IBMT only produced increased activity in the ventral ACC (Tang, 2017). These results suggest that the extent of neuroplasticity in the ACC following meditation may depend on the amount of training. Across the different time courses of IBMT the meditative state is always dominated by parasympathetic activity and accompanied by positive changes in the endocrine and immune systems. We believe that the change in autonomic system after training is critical for inducing the beneficial effects on stress reduction, such as decreased cortisol and increased immune function (Tang, 2017). The higher levels of cognitive function are likely due to improved regulation of emotion and thought through increased ACC connectivity (Tang et al., 2012, 2015). Further studies into this topic may allow us to delineate the importance of various biomarkers, such as increased brain activity or/and functional connectivity, improved axonal density, increased myelination, increased HF HRV, decreased SCR, and the relationships between these biomarkers and diverse behavioral consequences of meditation training. These behavioral changes include, but are not limited to, reduced stress, better immune function, increased positive moods and reduced negative moods, and better cognitive performance in attention, creativity, memory, and problem solving. Therefore ACC may serve as a "bridge" that connects and facilitates ANS and CNS interaction, and changes in these systems may be the result of the bodifulness and mindfulness techniques of meditation (Tang et al., 2019).

Individual differences in sympathetic and parasympathetic systems

Individual differences in self-control and learning ability exist and people respond to training and learning experience differently. People respond to meditation differently. These differences may derive from temperament, personality, brain, physiology, or genetic differences. Studies have suggested that genetic polymorphisms interact with experience to influence the success of training. Moreover, individual differences in lifestyle, life events, and trainer—trainee dynamics during meditation are likely to have substantial influence on training effects, though little is known about these influences (Tang, 2017). In a randomized study, we used short-term IBMT and an active control relaxation training to examine whether IBMT can improve creative performance and determine which individuals are most likely to benefit. We assessed mood, personality and creative performances before and after training. The results indicated that the IBMT group had significantly greater creative performance than the relaxation control, which is consistent with our previous studies (Ding, Tang, Tang, & Posner, 2014). A linear regression model showed that five predictors in pretests including depression, anger, fatigue, introversion × vigor, and emotional stability × vigor accounted for 57% of the variance in the change in creativity for IBMT group. This study suggested that mood and personality can be used to predict individual variations in the improvement of creative performance following mindfulness meditation (Ding, Tang, Deng, Tang, & Posner, 2015).

Appropriate attention or effort is crucial for meditation states. Does parasympathetic or sympathetic activity contribute to individual differences in attention and other cognitive tasks? One study combined HRV and ERP (ERP, a method that measures brain activity during tasks) to answer this question. The preliminary results showed that greater parasympathetic activity is a biomarker of increased selective attention and that heightened sympathetic activity also plays a role in more efficient attention-related processing, suggesting the importance of autonomic regulation in the study of individual differences in neurocognitive function. Additionally, parasympathetic and sympathetic activities may contribute to variability in brain function across a variety of cognitive tasks (Giuliano et al., 2018). However, future RCT studies with large sample sizes are warranted to confirm these results.

What is the relationship between personality traits and changes in EEG and HRV during meditation? During Zen meditation, fast theta power and slow alpha power in EEG increased in the frontal area, and increases in the HF HRV (as a parasympathetic index) and decreases in the LF HRV and LF/HF (as sympathetic indices) were also observed. EEG changes in alpha and theta power are consistent with a recent meta-analysis (Lomas et al., 2015). Further correlations were examined among these changes in terms of the percentage of change during meditation, as compared to a control state. The percentage of change in slow frontal alpha power (reflecting enhanced internalized attention) was negatively correlated with LF and LF/HF and was positively correlated with the novelty seeking score (which has been suggested to be associated with dopaminergic activity). The percentage of change in fast frontal theta power (reflecting enhanced mindfulness) was positively correlated with HF and also with the harm avoidance score (which has been suggested to be associated with serotonergic activity). Together these results suggest internalized attention and mindfulness during meditation may be characterized by different physiological properties including greater parasympathetic activity and more frontal alpha and theta power. Moreover, the relationship between EEG power and personality traits may imply that individual differences in personality could interact with physiological responses during meditation to influence the outcomes. In summary, capturing individual differences may serve to predict success in meditation training (Takahashi et al., 2005; Tang et al., 2015).

With regard to parasympathetic and sympathetic systems, studies have shown three different patterns: (1) Parasympathetic dominant pattern with greater HF HRV and lower SCR; (2) sympathetic dominant pattern with more LF HRV and higher SCR; and (3) a mixed pattern without dominance. In addition to the factors of personality and lifestyle, these different patterns across individuals may be due to the predisposition of genes and/or cultural effects (see Chapter 3: Cultural differences in meditation, and Chapter 4: Genetic association with meditation learning and practice (outcomes)). However, it is still possible to apply meditation to change these malleable indexes of parasympathetic and sympathetic activities such as HRV, SCR, and breathing patterns in frequency and amplitude. IBMT research has shown that trainees with sympathetic dominant patterns could change to parasympathetic dominant patterns following short-term practice (e.g., five 30 min sessions). But this may be a state (not a trait) change of the dominant pattern and we do not know if this change remains stable following stress or adverse events. However, a total of 5−10 h of IBMT (10−20 30-min sessions) could change brain structure in white and gray matter accompanied by behavioral changes such as emotion regulation, suggesting longer times of IBMT practice may induce trait change with respect to the dominant pattern. Neurophysiological changes in frontal theta and alpha power would also occur (Tang, 2017). For example, in one RCT study using short-term IBMT, we explored the relationship between brain activity (e.g., frontal midline theta power, electrodes Fz, FCz, and Cz source at the ACC) and physiological index (e.g., HF HRV). After five sessions of IBMT, correlations between Fz-theta and HF HRV, FCz-theta and HF HRV, and Cz-theta and HF HRV were all significantly positive. However, after 5-session relaxation training, there was no significant correlation between theta activity and HF HRV. While only the IBMT practice produced significant correlations, there were no significant differences between correlations for these two groups. Future studies should focus on examining individual differences in changes in sympathetic and parasympathetic systems and the CNS following mindfulness meditation (Tang, 2017; Tang et al., 2010).

Given the wide range of possible factors may contribute to individual differences in meditation practice and outcomes, developing a personalized approach using EEG-guided or HRV-guided techniques may be helpful. One study explored EEG-guided meditation and proposed quantitative EEG (qEEG) as a suitable tool to identify individual neurophysiological types based on cognitive performance and skills and brain activity. Results showed that qEEG screening may help developing a tailored meditation training program or protocol that helps maximize results and minimize risk of potential negative effects (Fingelkurts, Fingelkurts, & Kallio-Tamminen, 2015). In our work, we applied simpler HRV-guided or SCR-guided techniques which take shorter times to prepare and setup to help practitioners achieving meditative experiences and states. Our results suggested similar effects and the potential of using these techniques to guide meditation practices (Tang, 2009). These techniques are similar to genetic screening which could capture information on individual differences. For instance, even if individuals show weak parasympathetic system (e.g., lower HF HRV or higher SCR), which may be not ideal for meditation practice, meditation practices may still change these predisposition tendencies. Our research has shown rapid change to the parasympathetic system from the sympathetic system in novices following short-term IBMT practice, suggesting practice experiences can override predisposition tendencies. Moreover, if an individual's predisposition shows tendencies of strong parasympathetic system (e.g., higher HF HRV or lower SCR), these individuals may be able to meditate more smoothly and benefit more from practice. Future research is warranted to address the potential impact of individual differences in these physiological responses on meditation practices and outcomes.

Another important factor of the individual differences in sympathetic and parasympathetic systems is the dynamic changes between these two systems, since the extent of parasympathetic and sympathetic coordination is moderated by mental state, emotional context, stress levels, and other factors, suggesting that the autonomic coordination and regulation may vary dynamically within, and across, individuals (Gatzke-Kopp & Ram, 2018).

As we discussed in Chapter 3, Cultural differences in meditation, and Chapter 4, Genetic association with meditation learning and practice (outcomes), other physiological indexes such as stress (e.g., cortisol hormone) and immune, cardiovascular, and neuroendocrine responses may also influence parasympathetic and sympathetic systems. We will take stress as an example. Generally speaking, stress has two types: (1) somatic stress in the body, for example, a racing heart, indigestion, or the jitters; and (2) stress in mental cognitive processes, for example, worrisome thoughts that keep one up at night or that continually intrude one's attention during the day. Most importantly, the stress from body and mind often work together and interact. The stress-buffering hypothesis was proposed to explain how social support could help coping with stress and improve health outcomes (Cohen & Wills, 1985). However, inconsistencies in the literature exist. For example, two studies tested two models——the stress buffering hypothesis and the social identity model of identity change—to determine which model best explains the relationship between social connectedness, stress, and wellbeing. In study 1, participants considered the impact of moving to a different city versus receiving a serious health diagnosis. Study 2 examined the adjustment of visiting students to university over the course of their first semester. Both studies found limited evidence as predicted by the stress-buffering hypothesis; instead people who experienced a loss of social identities as a result of a stressor had a subsequent decline in well-being, consistent with the social identity model of identity change. These results seem to suggest that stressful life events are best conceptualized as identity transitions and that stress process and coping are much more complicated than previously thought (Praharso, Tear, & Cruwys, 2017). Based on the stress-buffering hypothesis, a mindfulness stress-buffering account was proposed to explain how meditation alters stress processes in the brain, which, in turn, reduces peripheral stress-response in the sympathetic-adrenal-medullary and hypothalamic–pituitary–adrenal (HPA) axes, as well as affecting health outcomes (Creswell & Lindsay, 2014). However, mindfulness meditation literature indicate that meditation reduces stress in the HPA axis through sympathetic and parasympathetic systems rather than the sympathetic system alone, while the parasympathetic system seems to respond faster to meditation practice (see reviews by Cahn & Polich, 2006; Lomas et al., 2015; Tang et al., 2015). Moreover, meditators in different meditative states or stages may also manifest in both systems and have distinct biological or physiological signatures, for example, HRV differences during early, middle, and advanced stages (Tang, 2017). These findings request an update of the hypothesis or model on how meditation helps individuals to handle stress and improve health outcomes through certain biological pathways.

Research also indicates that there is no a single or best way to erase cognitive and/or somatic anxiety. Not everyone would benefit from a body-focused method, just as meditation may not be the best way to fight stress for every individual. However, a body–mind technique that may help relax the body and calm the mind efficiently seems to be a promising choice and is worth the effort. It has been shown, for example, that a few 20-min sessions of IBMT can reduce stress as measured by the salivary hormone cortisol. When we calm down from stress, we are shifting our nervous system to a relaxed and calm state, which is known as parasympathetic dominance. In this state, our minds are more open and clearer, our HR slows down, breath deepens, and our muscles relax, allowing the potential release of rooted stress and tension (Tang, 2017). A recent meta-analysis suggested that HRV biofeedback training is associated with a large reduction in self-reported stress and anxiety (Goessl, Curtiss, & Hofmann, 2017). However, IBMT's regulation of HRV may work in a different mechanism compared to biofeedback training.

Meditation involves the interaction and harmony between both CNS (brain) and ANS (body) to support effective stress reduction and meditation practice. Compared to short-term meditation, long-term meditation may have lasting effects on sympathetic and parasympathetic systems. In a longitudinal RCT study we compared the effects of 10-year IBMT and physical exercise in groups of older adults (about 55 years old on average) and found that compared to the exercise group, the IBMT group had significantly higher ratings on dimensions of life quality. Parasympathetic activity indexed by SCR and HF HRV also showed more favorable outcomes in the IBMT group (see Fig. 5.1C). The horizontal axis indicates the four test sessions: 5-min resting with eyes open (baseline 1), two 10-min periods of resting with eyes closed (labeled as 1, 2), and 5-min resting with eyes open (baseline 2). The vertical axis indicates SCR change and percentage of change in normalized units of high-frequency HRV, respectively ($*p < .05$, $**p < .01$). However, the exercise group showed lower basal HR and greater chest respiratory amplitude (see Fig. 5.1D). The vertical axis indicates resting HR change on the left panel and percentage of change of chest BA on the right panel, respectively ($*p < .05$, $**p < .01$). Moreover, basal immune function, the sIgA level, was significantly higher and cortisol concentration was lower in the IBMT group. These results suggest more dominant parasympathetic activity following IBMT, but more sympathetic dominance following exercise.

Optimal performance—insights for meditation

Studies have long suggested that effort increases focus on the attentional target and increases distraction inhibition, and that this type of cognitive control enhances performance (Kahneman, 1973). However, more recent evidence has shown that a reduction in effortful control can also improve performance, such as in creativity, implicit learning, sensorimotor skills and mindfulness meditation (Bruya, 2010; Tang & Bruya, 2017), consistent with the findings and models in sport performance (Bertollo et al., 2016). Optimal performance often refers to an effortless and automatic, flow-like state of performance. Meditation regulates the focus of attention to optimally focus (balanced attention) on the core component of the action, avoiding too much attention that could be detrimental for elite performance. As a result, the meditator becomes skillful and enters into advanced stage of meditation. Balanced attention is a trained state that can optimize any particular attentional activity. One can exert minimal effort to maintain balanced attention, resulting in a large impact on performance in cognition, emotion, learning, training, health, and quality of life (Tang & Bruya, 2017; Tang & Posner, 2009). To optimize tasks that require high effort and explicit processing such as working memory, one can reallocate attentional resources, resulting in more efficiently focused attention and less effort. To optimize tasks that require low effort and implicit process such as creativity or sensorimotor skills, one can bring diffused attention to the task, resulting in more control and monitoring. Through balanced attention, different activities with different cognitive demands can be optimized with a balance of implicit and explicit processing, resulting in an appropriate level of attention and effort. Balanced attention has also been called the ''being state,', which can be achieved during deep meditation (Tang, 2017; Tang & Bruya, 2017; Tang et al., 2015).

There are two types of efforts in general——objective and subjective effort. Subjective effort is often based on self-reporting using questionnaires, whereas objective effort is evaluated by assessing task avoidance or physiological biomarkers. Given that mental effort is normally associated to bodily or somatic markers, some indexes such as SCR and HRV have been used in monitoring and measuring objective effort to validate the subjective self-reporting of effort. One study recruited healthy adults and one stroke patient with major ischemic lesion encompassing the left ACC. The patient had normal cognitive status as assessed by mini-mental state examination. While the healthy adults preserved both cognitive control abilities and conscious feeling of mental effort during a Stroop task (a task examining executive attention and cognitive control), the patient preserved control abilities but lost the ability to experience and report a feeling of mental effort. When a healthy subject experiences mental effort, his or her body has detectable reactions affecting HR, blood pressure, breathing rate, pupillary diameter, or SCR (Naccache et al., 2005). Clinical neuropsychology and brain imaging studies suggest a crucial role of the ACC in the generation of these somatic signals. Compared to healthy controls, the tight coupling between cognitive difficulty (mental effort) and somatic response SCR was absent in the patient, indicating that the lack of mental effort was associated with an impairment of the somatic marker SCR (Naccache et al., 2005). Therefore SCR is a reliable somatic marker of mental effort and cognitive difficulty and may be generated by the ACC.

Previous research has suggested that effortless attention and control lead to optimal performance including meditation practice (Bruya, 2010; Tang & Bruya, 2017; Tang et al., 2015). Theta and alpha power consistently show up during meditation (Lomas et al., 2015). Converging evidences in sports also indicate that the frontal midline theta is a biomarker of optimal–automatic performance (Bertollo et al., 2016), which is in line with our series of IBMT work indicating that frontal midline theta is a biomarker of the deep meditation state. Moreover, our human and animal research also showed that the frontal midline theta power induces brain structural changes that support behavioral changes such as emotion regulation (Tang, 2017; Tang et al., 2019). In addition to frontal midline theta, are there other biomarkers of deep meditation state? Our research has indicated that more parasympathetic activity is involved in deep meditation and the SCR and HRV are major physiological indexes of meditative states (Tang, 2017; Tang et al., 2015, 2019). During brief IBMT training, frontal midline ACC theta was correlated with high-frequency HRV, suggesting control by the ACC over parasympathetic activity (Tang, 2017; Tang et al., 2010). In other words, brief IBMT induces better regulation of the ANS through a midline ACC brain system, suggesting the interaction and coordination of body and mind following IBMT, a form of mindfulness meditation that optimizes activities for maximum self-control, attention, and efficiency with minimal effort (Tang, 2017). Other studies have also shown that parasympathetic activity is associated with the flow-state (de Manzano, Theorell, Harmat, & Ullén, 2010; Keller, Bless, Blomann, & Kleinbohl, 2011), an example of balanced attention in which high control is achieved with low subjective mental effort

(Bruya, 2010). These findings are important because a feeling of mental effort often triggers action avoidance and leads to quitting in training programs such as meditation, physical exercise, and other healthy behavior and lifestyles. If we could achieve balanced attention (via IBMT or other means) with less mental effort, this experience could potentially motivate us to stick to our goal and continue to practice. We call this mechanism the "parasympathetic-mind—body interaction.'. In summary, growing empirical evidence indicates that parasympathetic-mind—body interaction often triggers optimal performance and is one possible mechanism for optimizing performance. Parasympathetic-mind—body interaction could have a large impact on positive emotion, health benefits, quality of life, and self-growth (Bruya, 2010; Tang, 2017; Tang & Posner, 2009; Tang et al., 2012). It should be noted that HRV as an index of mental effort is still in debate, possibly because of the complicated relationships among the HF, LF, and LF/HF components of HRV.

Parasympathetic activity in brain—gut interaction and disorders

Not only do HF HRV (vagus tone) serve as a biomarker of meditative states, but also indexes the brain—gut interaction in many disorders. For example, the vagus nerve is the longest nerve and a major component of the parasympathetic system, which oversees many crucial bodily functions, including control of HR, digestion, mood, and immune response. There is an equilibrium between the parasympathetic and sympathetic systems that is responsible for the maintenance of homeostasis. If an imbalance of these two systems occurs, various pathologic conditions will happen. Given that the vagus nerve is a mixed nerve with 80% afferent and 20% efferent fibers, it is also a key component of the neuroimmune and brain—gut axes through a bidirectional communication between the brain and the gastrointestinal tract. The vagus nerve plays a dual anti-inflammatory role through either vagal afferents targeting the HPA axis, or vagal efferents targeting the cholinergic anti-inflammatory pathway (Bonaz, Sinniger, & Pellissier, 2016). The vagus nerve and sympathetic system act in synergy to inhibit the release of tumor necrosis factor alpha. Because of its anti-inflammatory effect, the vagus nerve may be a therapeutic target in the treatment of chronic inflammatory disorders such as inflammatory bowel diseases and irritable bowel syndrome. Diverse therapeutic strategies could potentially target the anti-inflammatory properties of the vagus nerve, such as medications, enteral nutrition, complementary medicines (e.g., meditation), physical exercise, or vagus nerve stimulation (VNS). VNS is one of the alternative treatments for drug-resistant epilepsy and depression and could be used as a nonpharmacological therapy to treat inflammatory disorders, which are characterized by a blunted autonomic balance with a decreased vagal tone. The vagus nerve seems to be a modulator of the brain-gut axis in psychiatric and inflammatory disorders (Bonaz et al., 2016; Bonaz, Sinniger, & Pellissier, 2017).

The vagus nerve establishes one of the connections between the brain and the gastrointestinal tract and sends information about the state of the inner organs to the brain via afferent fibers. To be a potential target in treating inflammatory bowel disease, the vagus nerve is also a target in treating psychiatric disorders. Preliminary evidence suggests that VNS is a promising treatment for treatment-resistant mood disorders including depression, anxiety, PTSD, and others. VNS increases the vagal tone and inhibits cytokine production, which is an important mechanism of resiliency. Moreover, stimulating vagal afferent fibers in the gut influences monoaminergic brain systems in the brain stem, which play crucial roles in major psychiatric disorders. In the same vein, preliminary evidence suggests that gut bacteria have a beneficial effect on mood disorders, partly by affecting the vagus nerve activity. The vagal tone is correlated with the capacity to regulate stress responses and can be affected by breathing, so meditation could potentially increase the vagal tone, thus contributing to resilience, reduction of symptoms, and treatment of mental disorders (Breit, Kupferberg, Rogler, & Hasler, 2018). Meditation techniques that can change and regulate the vagus nerve (vagus tone) through breathing, may hold promises for clinical applications. However, different breathing patterns can induce greater sympathetic or parasympathetic activities, but only parasympathetic activity contributes to vagus tone. Therefore the selection of meditation technique or breathing styles is crucial (Tang, 2017). In summary, accumulating data indicate how variations and changes in the composition of the gut microbiota influence physiology and contribute to diseases. Furthermore, evidence also suggests that the gut microbiota communicates with the brain possibly through neural, endocrine, and immune pathways, thereby influencing brain function and behavior. A microbiota-gut-brain axis suggests that modulation of the gut microbiota may help develop novel therapeutics for complex disorders such as mood disorders, inflammation, pain, and obesity (Cryan & Dinan, 2012). Taken together, more evidence indicate the important role of parasympathetic system in meditation and the latest research also suggests the key role of parasympathetic activity in brain−gut interaction and disorders. Therefore meditation may also work on brain−gut interaction and help prevent and treat disorders.

Practice tips

1. We could apply the autonomic markers such as HRV and SCR to help meditation learning and practice. First, we could use premeditation HRV or SCR as a screening index to obtain individual-specific information about participants, and then select the proper meditation technique(s) or design corresponding training protocols for the individual. Second,

we could give feedback to the participants based on HRV or SCR index during meditation.

2. Every day, our physiological rhythm (sympathetic and parasympathetic activities) change over time, therefore we could find the optimal time of meditation practice based on physiology fluctuation such as greater HF HRV or lower SCR.

3. New evidence indicates that a reduction in effortful control improves performance in different tasks and activities. It is possible that the benefits of meditation could be enhanced by reduction in effortful control. To achieve optimal performance in meditation we could provide individuals with their SCR results and help them better control the amount of effort and have an appropriate and optimal amount of effort when they practice. This would help them enter meditative states more easily.

References

Babo-Rebelo, M., Richter, C. G., & Tallon-Baudry, C. (2016). Neural responses to heartbeats in the default network encode the self in spontaneous thoughts. *The Journal of Neuroscience*, *36*(30), 7829−7840.

Bertollo, M., di Fronso, S., Conforto, S., Schmid, M., Bortoli, L., Comani, S., & Robazza, C. (2016). Proficient brain for optimal performance: The MAP model perspective. *PeerJ*, *4*, e2082.

Billman, G. E. (2013). The LF/HF ratio does not accurately measure cardiac sympatho-vagal balance. *Frontiers in Physiology*, *4*, 26.

Bonaz, B., Sinniger, V., & Pellissier, S. (2016). Vagal tone: Effects on sensitivity, motility, and inflammation. *Neurogastroenterology and Motility*, *28*(4), 455−462.

Bonaz, B., Sinniger, V., & Pellissier, S. (2017). The vagus nerve in the neuro-immune axis: Implications in the pathology of the gastrointestinal tract. *Frontiers in Immunology*, *8*, 1452.

Breit, S., Kupferberg, A., Rogler, G., & Hasler, G. (2018). Vagus nerve as modulator of the brain-gut axis in psychiatric and inflammatory disorders. *Frontiers in Psychiatry*, *9*, 44.

Bruya, B. (2010). *Effortless attention: A new perspective in the cognitive science of attention and action*. Boston, MA: MIT Press.

Bush, G., Luu, P., & Posner, M. I. (2000). Cognitive and emotional influences in anterior cingulate cortex. *Trends in Cognitive Sciences*, *4*(6), 215−222.

Cahn, B. R., & Polich, J. (2006). Meditation states and traits: EEG, ERP, and neuroimaging studies. *Psychological Bulletin*, *132*(2), 180.

Cohen, S. E., & Wills, T. A. (1985). Stress, social support, and the buffering hypothesis. *Psychological Bulletin*, *98*, 310−357.

Creswell, J. D., & Lindsay, E. K. (2014). How does mindfulness training affect health? A mindfulness stress buffering account. *Current Directions in Psychological Science*, *23*(6), 401−407.

Critchley, H. D. (2004). The human cortex responds to an interoceptive challenge. *Proceedings of the National Academy of Sciences of the United States of America*, *101*(17), 6333−6334.

Cryan, J. F., & Dinan, T. G. (2012). Mind-altering microorganisms: The impact of the gut microbiota on brain and behaviour. *Nature Reviews Neuroscience*, *13*(10), 701−712.

de Manzano, O., Theorell, T., Harmat, L., & Ullén, F. (2010). The psychophysiology of flow during piano playing. *Emotion (Washington, D.C.)*, *10*, 301−311.

Dijkstra, K., Kaschak, M. P., & Zwaan, R. A. (2007). Body posture facilitates retrieval of auto-biographical memories. *Cognition, 102,* 139−149.

Ding, X., Tang, Y. Y., Tang, R., & Posner, M. I. (2014). Improving creativity performance by short-term meditation. *Behavioral and Brain Functions, 10,* 9.

Ding, X., Tang, Y. Y., Deng, Y., Tang, R., & Posner, M. I. (2015). Mood and personality predict improvement in creativity due to meditation training. *Learning and Individual Differences, 37,* 217−221.

Fingelkurts, A. A., Fingelkurts, A. A., & Kallio-Tamminen, T. (2015). EEG-guided meditation: A personalized approach. *Journal of Physiology, Paris, 109*(4-6), 180−190.

Gatzke-Kopp, L., & Ram, N. (2018). Developmental dynamics of autonomic function in child-hood. *Psychophysiology, 55*(11), e13218.

Gerritsen, R. J. S., & Band, G. P. H. (2018). Breath of life: The respiratory vagal stimulation model of contemplative activity. *Frontiers in Human Neuroscience, 12,* 397.

Giuliano, R. J., Karns, C. M., Bell, T. A., Petersen, S., Skowron, E. A., Neville, H. J., & Pakulak, E. (2018). Parasympathetic and sympathetic activity are associated with individual differences in neural indices of selective attention in adults. *Psychophysiology, 55*(8), e13079.

Gizowski, C., & Bourque, C. W. (2018). The neural basis of homeostatic and anticipatory thirst. *Nature Reviews Nephrology, 14*(1), 11−25.

Goessl, V. C., Curtiss, J. E., & Hofmann, S. G. (2017). The effect of heart ráte variability biofeedback training on stress and anxiety: A meta-analysis. *Psychological Medicine, 47*(15), 2578−2586.

Hennig, J., Friebe, J., Ryl, I., Krämer, B., Böttcher, J., & Netter, P. (2000). Upright posture influ-ences salivary cortisol. *Psychoneuroendocrinology, 25,* 69−83.

Huang, L., Galinsky, A. D., Gruenfeld, D. H., & Guillory, L. E. (2011). Powerful postures versus powerful roles: Which is the proximate correlate of thought and behavior? *Psychological Science, 22,* 95−102.

Jiang, H., He, B., Guo, X., Wang, X., Guo, M., Wang, Z., . . . Cui, D. (2019). Brain-heart inter-actions underlying traditional tibetan buddhist meditation. *Cerebral Cortex.* Available from https://doi.org/10.1093/cercor/bhz095.

Kahneman, D. (1973). *Attention and effort.* Englewood Cliffs, NJ: Prentice Hall.

Keller, J., Bless, H., Blomann, F., & Kleinbohl, D. (2011). Physiological aspects of flow experi-ences: Skills-demand-compatibility effects on heart rate variability and salivary control. *Journal of Experimental Social Psychology, 47,* 849−852.

Kerr, C. E., Sacchet, M. D., Lazar, S. W., Moore, C. I., & Jones, S. R. (2013). Mindfulness starts with the body: Somatosensory attention and top-down modulation of cortical alpha rhythms in mindfulness meditation. *Frontiers in Human Neuroscience, 7,* 12.

Lomas, T., Ivtzan, I., & Fu, C. H. (2015). A systematic review of the neurophysiology of mind-fulness on EEG oscillations. *Neurosci Biobehav Rev, 57,* 401−410.

Lutz, A., Greischar, L. L., Perlman, D. M., & Davidson, R. J. (2009). BOLD signal in insula is differentially related to cardiac function during compassion meditation in experts vs. novices. *NeuroImage, 47*(3), 1038−1046.

Naccache, L., Dehaene, S., Cohen, L., Habert, M. O., Guichart-Gomez, E., Galanaud, D., et al. (2005). Effortless control: executive attention and conscious feeling of mental effort are dis-sociable. *Neuropsychologia, 43,* 1318−1328.

Niedenthal, P. M. (2007). Embodying emotion. *Science, 316,* 1002−1005.

Park, G., & Thayer, J. F. (2014). From the heart to the mind: Cardiac vagal tone modulates top-down and bottom-up visual perception and attention to emotional stimuli. *Frontiers in Psychology, 5,* 278.

Pocock, G. (2006). *Human physiology* (3rd ed.). Oxford University Press.

Posner, M. I., & Rothbart, M. K. (2006). *Educating the human brain.* APA Press.

Posner, M. I., Sheese, B., Rothbart, M., & Tang, Y. Y. (2007). The anterior cingulate gyrus and the mechanism of self-regulation. *Cognitive, Affective, & Behavioral Neuroscience, 7*(4), 391–395.

Praharso, N. F., Tear, M. J., & Cruwys, T. (2017). Stressful life transitions and wellbeing: A comparison of the stress buffering hypothesis and the social identity model of identity change. *Psychiatry Research, 247,* 265–275.

Quintana, D. S., & Heathers, J. A. J. (2014). Considerations in the assessment of heart rate variability in biobehavioral research. *Frontiers in Psychology, 5,* 805.

Shackman, A. J., Salomons, T. V., Slagter, H. A., Fox, A. S., Winter, J. J., & Davidson, R. J. (2011). The integration of negative affect, pain and cognitive control in the cingulate cortex. *Nature Reviews Neuroscience, 12*(3), 154–167.

Takahashi, T., Murata, T., Hamada, T., Omori, M., Kosaka, H., Kikuchi, M., ... Wada, Y. (2005). Changes in EEG and autonomic nervous activity during meditation and their association with personality traits. *International Journal of Psychophysiology, 55,* 199–207.

Takatani, T., Takahashi, Y., Yoshida, R., Imai, R., Uchiike, T., Yamazaki, M., ... Fujimoto, S. (2018). Relationship between frequency spectrum of heart rate variability and autonomic nervous activities during sleep in newborns. *Brain and Development, 40*(3), 165–171.

Tang, Y. Y. (2017). *The neuroscience of mindfulness meditation: How the body and mind work together to change our behavior?* London: Springer Nature.

Tang, Y. Y., & Bruya, B. (2017). Mechanisms of mind-body interaction and optimal performance. *Frontiers in Psychology, 8,* 647.

Tang, Y. Y., & Posner, M. I. (2009). Attention training and attention state training. *Trends in Cognitive Sciences, 13*(5), 222–227.

Tang, Y. Y., Ma, Y., Fan, Y., Feng, H., Wang, J., ... Fan, M. (2009). Central and autonomic nervous system interaction is altered by short term meditation. *Proceedings of the National Academy of Sciences, USA, 106*(22), 8865–8870.

Tang, Y. Y., Rothbart, M. K., & Posner, M. I. (2012). Neural correlates of establishing, maintaining and switching brain states. *Trends in Cognitive Sciences, 16,* 330–337.

Tang, Y. Y., & Tang, R. (2015). Mindfulness: Mechanism and application. In A. W. Toga (Ed.), *Brain mapping: An encyclopedic reference* (3, pp. 59–64). Elsevier: Academic Press.

Tang, Y. Y., Ma, Y., Wang, J., Fan, Y., Feng, S., Lu, Q., & Posner, M. I. (2007). Short term meditation training improves attention and self regulation. *Proceedings of the National Academy of Sciences, USA, 104*(43), 17152–17156.

Tang, Y. Y., Holzel, B. K., & Posner, M. I. (2015). The neuroscience of mindfulness meditation. *Nature Reviews Neuroscience, 16,* 213–225.

Tang, Y. Y., Tang, R., & Gross, J. J. (2019). Promoting emotional well-being through an evidence-based mindfulness training program. *Frontiers in Human Neuroscience, 13,* 237.

Tyagi, A., & Cohen, M. (2016). Yoga and heart rate variability: A comprehensive review of the literature. *International Journal of Yoga, 9*(2), 97–113.

Wallace, R. K. (1970). Physiological effects of transcendental meditation. *Science, 167,* 1751–1754.

Chapter 6

Brain regions and networks in meditation

The neuroscientific investigation of meditation was not significant until neuroimaging techniques and methodologies started to flourish in the fields of cognitive neuroscience and psychology. With the help of neuroimaging technologies we are able to examine brain activations and blood flow while individuals perform certain tasks or are in specific mental states. However, it is worth taking the time to discuss why we are interested in the brain when our goal is to understand meditation-related benefits in behavior. In the beginning of meditation research a considerable amount of effort was devoted to understanding behavioral changes after meditation experiences. This focus on behavior is not surprising because before the value of cognitive neuroscience was widely appreciated, most psychologists were primarily interested in behavior and not so much in the brain. One question that was often asked among psychologists is why we should care about the brain if we already know that under X condition, we would observe Y behavior, or providing X intervention, we would see improvement in Y behavior. In other words, do we need to study the brain to understand the mind? The question of whether studying the brain is necessary for understanding the mind was widely debated by psychologists who focused on examining only human behavior and cognitive processes. However, as opponents of brain research would soon realize, understanding the mechanisms underlying observable behavior is equally as important as observing behavior and changes in behavior.

The converse question was asked by neuroscientists who focused on studying neurons and the brain at molecular, cellular, and system levels, but did not consider how these findings may translate into behavior and higher-level cognitive processes. Fortunately, both sides seemed to gradually arrive at the consensus that studying the implementation of behavior is important (Krakauer, Ghazanfar, Gomez-Marin, MacIver, & Poeppel, 2017; Wager, 2006). This means that neuroscience needs psychology to explain observed physiological and neural phenomenon in relation to mind and behavior, and psychology needs neuroscience to understand how behavior and cognitive processes are represented in the brain and implemented and carried out in practice. Ultimately, both questions are interesting and equally important for

The Neuroscience of Meditation. DOI: https://doi.org/10.1016/B978-0-12-818266-6.00007-1
109

obtaining a holistic and comprehensive understanding of the human mind and brain and for further advancing our scientific knowledge and clinical applications.

In addition to understanding the implementation of behavior, studying the brain has several pivotal implications for fleshing out our knowledge of human cognitive processes and mental states. Typically, psychologists would run into the question of similarity and distinctiveness between various cognitive processes. Yet task performance and observation of people performing different cognitive tasks would not provide us with clues regarding the underlying differences or similarities between each cognitive process (Wager, 2006). For example, task performance of memory and attention are likely to be similar, but this does not tell us anything about the underlying processes that are activated when completing either of the two tasks, except that it tells us attention and memory are correlated in some ways. This issue of specificity with respect to cognitive processes would remain elusive if we have no way of directly measuring the dynamics of these somewhat hidden processes. Even if we propose tentative hypotheses and theories of how cognitive processes are executed and different or alike, we still lack concrete and convincing empirical evidence to support our claims.

Neuroimaging techniques such as functional magnetic resonance imaging (fMRI) can tackle this issue of specificity. In particular, fMRI has high spatial resolution and can reveal regions in the brain associated with different cognitive processes and demonstrate differences and similarities in brain activation while performing various cognitive tasks. For instance, if we see distinct brain regions activated separately for attention and memory tasks, we are able to make inferences that these two aspects of cognitive function may involve unique underlying mental processes. This objective measure of brain activation advances us further into the hidden processes that support our daily functioning. To some extent knowing the neural basis of cognitive processes and behavior offers a glimpse into the physical constraints put on our abilities and enables us to have a deeper insight with respect to the connection between mind and behavior. Therefore the rise of cognitive neuroscience and its methodologies greatly inspired neuroimaging studies of meditation and meditation-based interventions.

Motivated by this need for a thorough understanding of the underlying mechanisms associated with meditation-induced changes in psychological well-being, cognitive function, and physiology, researchers have undertaken investigations from interesting angles. There are two types of fMRI studies conducted for examining brain functional activation and connectivity, namely (1) task fMRI and (2) resting-state fMRI. Task fMRI was originally the most widely employed approach for studying the brain. This approach involves designing a task for participants to complete while they lie in an magnetic resonance imaging (MRI) scanner. The major hypothesis behind this type of study was that any observed brain activation during the fMRI scanner would be evoked by tasks and task stimuli. In the data analysis of task fMRI the onset and offset time of each task stimuli are recorded and later modeled to identify task-evoked brain activation. For instance, if one were to complete a memory task and make button presses for responses, brain regions that are activated during the tasks, especially when task stimuli are presented and responses are made by participants, would be related to task performance. However, identifying the role of each brain region would often require careful experimental manipulation and interpretation based on prior findings. The actual research process is not as straightforward as what we describe here, but we mainly sought to briefly illustrate what task fMRI entails.

The second type of fMRI study is resting-state fMRI. The resting-state is when no task is given to participants and they are simply asked to lie still in the MRI scanner and not think of anything in particular. This design often involves participants staring at a fixation cross that is shown in the center of the screen or having their eyes closed during the entire fMRI scan. This imaging protocol is particular helpful for developmental research on infants, young children, and other populations (see Chapter 7: Meditation over the lifespan). Studies have shown some subtle differences in the brain between these two resting conditions, but scientific communities have yet to reach a consensus on how to best acquire resting-state fMRI data (Patriat et al., 2013; Zuo & Xing, 2014). Different from task fMRI data, resting-state fMRI data are typically analyzed through computing functional connectivity. Functional connectivity refers to a statistical dependence between two brain regions across time and obtained through calculating the temporal correlations (Pearson's r) between the time-series of these two brain regions (Fox & Raichle, 2007). Resting-state signals are thought of as spontaneous neuronal activity within the brain because there is an absence of input from tasks to evoke a response from the brain. The lack of task input actually helps the investigations of the intrinsic architecture and functional organization of the brain, as regions that are highly correlated across time with each other in signals during rest are also likely to be functionally related. Indeed, through computing functional connectivity in the resting state, a growing body of literature has characterized and identified distinct brain functional networks across the brain, each related to somewhat different aspects of human cognition and behavior (Bullmore & Sporns, 2009; Fox et al., 2005; Power et al., 2011; Schaefer et al., 2017; Tang, Tang, Tang, & Lewis-Peacock, 2017).

Not only are brain functional activation and connectivity relevant for behavior and cognitive function, brain structures and their anatomical (i.e., physical) connectivity are also highly important for these critical processes. Structural MRI is another kind of neuroimaging technique that also includes two types of study. The first is structural MRI studies that measure the morphometry of the brain that involve indexes such as volume and cortical thickness of different brain regions and anatomical structures. Typically, for structural studies that focus on volume and cortical thickness, researchers would examine the gray matter of the brain (mainly the outer layer of the cerebral cortex, made up of neuronal cell bodies) by comparing the structural volume or cortical thickness of a region in the brain between two different groups (e.g., healthy adults vs depressed adults, or meditators vs nonmeditators). If we find patients with depression have a lower amygdala (a region responsible for emotion-related processes) volume than healthy individuals, then this could suggest that emotional problems associated with depression may manifest as reduced amygdala volume in the brain; if we find greater insula (a region associated with interoceptive awareness) volume in meditators compared to nonmeditators, this could suggest that meditation induces insula structural plasticity. The second type of structural MRI studies examines structural connectivity of the brain where the brain's white matter is the focus of investigation. White matter contains fiber bundles or tracts situated underneath the gray matter and are made up of highly myelinated axons that connect different regions of the brain. For instance, the left hemisphere of the brain is connected to the right hemisphere of the brain through the corpus callosum, which are white matter fibers that allow interhemispheric communication. One thing to note is that although white matter tracts and fibers connect different brain regions and structures, the functional connectivity between regions and structures are not necessarily constrained by this structural connectivity via white matter. In other words, two regions may be functionally correlated without being physically connected through white-matter tracts. Many mental processes including meditation show this pattern. With regard to white matter, common indexes of structural connectivity include directionality measures of white-matter microstructures such as fractional anisotropy and axonal diffusivity (Jones, Knösche, & Turner, 2013), which can be used to draw inferences regarding white-matter integrity and are also typically compared between two different groups (e.g., healthy vs depressed adults).

Although the primary focus of this chapter is on fMRI studies of meditation, it is worth noting that meditation studies using electroencephalography (EEG) to assess electrical activity of the brain have also been hugely informative with regard to revealing brain mechanisms and states relevant for meditation. As a noninvasive tool, EEG has good temporal resolution and is realized through placing small electrodes on the scalp of the brain, which would then detect electroencephalographic activity and fluctuations arising from the brain. Notably, these EEG activities exhibit oscillations at various frequencies, which can be categorized into bands: delta (less than 4 Hz), theta (4-7 Hz), alpha (8−15), beta (16−31 Hz), and gamma (greater than 32 Hz). Each of these frequency bands relates to different mental states and activity and have shown abnormalities in pathological populations. With regard to meditation, different frequency bands have been found to change during and following meditation, but alpha and theta bands are consistently found during and following meditation. However, it has been inconclusive as to the directions of these changes, since some showed increased in power, while others found decreases in power (Cahn & Polich, 2006; Lomas, Ivtzan, & Fu, 2015). However, compared to the fMRI technique, EEG suffers from relatively poor spatial resolution since the electrical activity picked up by electrodes are usually contributed by electrical events coming from multiple layers underneath the scalp, resulting in a mixture of sources of signals that are often blurred by these resistive layers (Burle et al., 2015). This makes it challenging to precisely localize the source of EEG signals. Although our focus of discussion in on fMRI studies of meditation, interested readers are encouraged to refer to reviews of EEG studies on meditation (see Cahn & Polich, 2006; Lomas et al., 2015).

The human brain is a fascinating and complex system that supports our daily functioning. Many brain regions have been determined to specialize in some of these critical functions, such as the visual cortex being responsible for vision and the motor cortex being responsible for movement. Interestingly, neuroimaging studies have even localized representations of body parts such as fingers, nose, and toes in the somatosensory cortex of the brain. However, unlike these findings that demonstrate clear cortical mapping of certain aspects of behavior and body parts, higher-level cognitive processes and mental states including the meditative state are far more complicated to be mapped exactly onto specific brain regions or cortices. The truth is that most of the time we identify neural correlates of cognitive functions and mental states, but we cannot pinpoint one region and claim that it is solely responsible for a specific function. In fact, many of these neural correlates or regions are located in the so-called "association cortex," which is likely to support multiple functions and higher-order mental processes. Therefore for the discussion of neuroimaging studies of meditation, readers should keep in mind that multiple brain regions are involved in meditation and it is very likely that they together contribute to achieving and maintaining a meditative state, rather than isolated brain regions that solely responsible for the state. Additionally, there is a wide range of meditation techniques taught in standardized meditation programs and interventions such as concentration, relaxation, body scan, and mindfulness. Whenever researchers assess intervention outcomes following this kind of standardized program, it is difficult to know precisely which technique contributes to what kind of change in brain and behavior, or if one technique has more impact on the outcomes than the other techniques (Tang, Hölzel, & Posner, 2015; Tang, Lu, Feng, Tang, & Posner, 2015). However, there have been some studies specifically examining the neural correlates of individual techniques, which we will also discuss along with the studies on standardized meditation programs.

Structural magnetic resonance imaging studies of meditation

We begin our discussion with structural studies of meditation focusing on gray matter morphometry, since these are straightforward investigations that compare differences in morphometric indexes between different groups (e.g., healthy adults vs experienced meditators, cross-sectional) or within the same group across different time points (before vs postintervention; longitudinal). Before we move onto a review of structural neuroimaging studies, an introduction to these brain morphometric indexes is needed for interpreting the significance of such findings. Given the wide array of tools available for brain structural investigations, there are often many structural indexes of gray matter available for assessing meditation-related changes in the brain. Knowing the differences between these indexes is necessary for understanding what kind of change took place in the brain following short-term or long-term meditation practice. In a review of structural neuroimaging studies of meditation (Fox et al., 2014), four major commonly used structural indexes of gray matter in meditation literature are summarized. The first is volume, which basically measures the volume of a brain region where a higher volume typically implies larger size, yet it is less clear what kind of neuronal changes are involved (i.e., more neurons, thicker cortex). The second index is density or concentration, which refers to the proportion of gray matter to other tissue type within a region or structure and should not be confused with cell packaging density (Mechelli, Price, Friston, & Ashburner, 2005). The third common index is cortical thickness, which is a straightforward measure of the thickness of a brain region where increases in thickness are usually linked to greater number of neurons/glia in a region. The fourth index is gyrification, which assesses the degree of folding of cortical surface on a point-by-point basis where higher gyrification is related to an increase in cortical surface area per unit volume (Fox et al., 2014).

As we can see from the descriptions of these indexes, they are all related yet different in some ways. In some studies we see more regions that are detected to improve or differ between groups when using one index versus another index in people who practice the same type of meditation. This speaks to the fact that the underlying neuronal and cellular mechanisms of changes in these indexes are different and we have yet to reach a conclusive agreement on what these mechanisms are. Furthermore, there are many possibilities and sources of contribution at the neuronal level that may lead to increases or decreases in these indexes, thus we should always be careful in our interpretation and should refrain from making definitive conclusions regarding the cellular mechanisms of changes in these structural indexes. Another misconception is also worth noting—more is better. Increases in structural indexes are often interpreted as improvement in function and behavior. This interpretation is troublesome as there are clearly cases where more is not better (e.g., neural pruning is abnormal in autistic individuals). Structural changes in meditation may also show a similar pattern. The bottom line is that when making an interpretation about structural changes in the brain, one should always be careful about drawing a one-to-one connection between brain and behavior based only on structural results, unless there is strong empirical evidence supporting a direct association between changes in structural indexes and changes in some behavior indexes.

Cortical thickness

The first structural study on brain gray matter can be traced back to almost 15 years ago when Lazar et al. (2005) examined the cortical thickness of experienced Buddhist Insight meditation practitioners and compared it with that of matched healthy controls who did not have any meditation experience. According to the cross-sectional study, Insight meditation is a form of meditation that does not involve reciting mantras or chanting as part of the practice. Instead, the primary technique of Insight meditation involved promoting mindfulness through cultivating attention and awareness of present-moment experiences (Lazar et al., 2005). The goal was to foster a nonjudgmental awareness and maintain attention focus to internal and external sensations, which is consistent with the secular definition of meditation practices. Results showed that right anterior insula as well as middle and superior frontal sulci showed greater cortical thickness in meditators than in controls. This study employed a cross-sectional research design in which all participants were assessed at one time point and compared as different groups of populations. Most structural MRI studies to date have utilized this type of design, perhaps due to the fact that structural changes are rarely found among people who receive short-term meditation intervention, making baseline and postintervention comparisons less likely to yield a significant effect. Another study of cortical thickness focused on experienced Zen meditators and found that they had greater cortical thickness in the right dorsal anterior cingulate cortex (ACC) and bilateral secondary somatosensory cortex compared to healthy controls who did not have meditation experience (Grant, Courtemanche, Duerden, Duncan, & Rainville, 2010). Zen meditation is generally characterized by concentrating on breathing, noticing episodes of mind-wandering, and maintaining attention focus with an open attitude and a straight posture. The attentional aspect of Zen meditation seems to be analogous to the description of Insight meditation. The study also showed that years of meditation experience was positively related to cortical thickness in the dorsal ACC, suggesting that more practice may likely induce further enhancement. In a separate study on Zen meditators and healthy controls, Grant et al. (2013) showed that the left supramarginal gyrus, left superior parietal lobule, and left superior frontal gyrus had greater cortical thickness in meditators than in controls, inconsistent with prior results. Future studies are warranted to examine the underlying reasons of these different findings. For instance, different stages (e.g., early or advanced) of meditators could contribute to these differences (Tang, Hölzel et al., 2015; Tang, Lu et al., 2015). Taken together, there is a fair number of regions in the brain that have shown greater cortical thickness in experienced meditators with long-term meditation experience when compared to those without any meditation experience.

Volume

In terms of gray-matter volume, one study found that the left putamen suffered less from age-related decline in Zen meditators than in age-matched controls (Pagnoni & Cekic, 2007), suggesting that meditation practice may be beneficial for slowing down normal aging. Likewise, volumetric differences were also found in right orbitofrontal cortex, right thalamus, and left inferior temporal gyrus for long-term meditators (it is unknown what kind of meditation was practiced by these people) when compared against healthy controls (Luders, Toga, Lepore, & Gaser, 2009). Furthermore, bilateral hippocampal volumes were found to be greater in long-term meditators (a mix of different meditation techniques) than healthy controls (Luders et al., 2013). Interestingly, one study of short-term meditation intervention using 8-week mindfulness-based stress reduction (MBSR) also found volumetric improvement, such that there was a greater left caudate volume at postintervention for novices who underwent the intervention than those in the waitlist control group (Farb, Segal, & Anderson, 2012). The MBSR program is a multifaceted package where a mixture of different techniques is taught to individuals during the intervention program, which is different from the more specific types of meditation practices we have discussed so far in this chapter. Interestingly, a study of experienced meditators who practiced loving kindness meditation showed greater gray matter volume in the right angular gyrus, right parahippocampal gyrus, and left inferior temporal gyrus than matched controls (Leung et al., 2012). Loving kindness meditation is one of the major techniques taught in standardized meditation programs and mainly involves fostering a sense of friendliness, warmth, and kindness toward oneself and others through thoughts, words, and imagery. It should be noted that all these studies used cross-sectional design which precludes causality. Furthermore, some studies did not clarify the meditation techniques used, which may create confusion. Nevertheless, gray matter seems to be influenced by both short-term and long-term meditation experiences.

Concentration/density

Switching to studies focusing on gray matter concentration or density as the primary index, we would also notice some of these common brain regions. Similar to Lazar et al. (2005), another study was conducted with experienced Insight meditation practitioners and healthy controls. This time the results indicated that meditators had greater gray matter density in left inferior temporal gyrus (same region was found in volumetric index), right insula (a region that was found before using cortical thickness as index), and right hippocampus (Hölzel et al., 2007). Based on this result, it seems that the left inferior frontal gyrus and right insula have been consistently showing up across different techniques of meditation and structural indexes. A different study on individuals who underwent an 8-week MBSR program showed decreases in amygdala density, a structure that is linked to emotion and stress response. Moreover, this decrease in density is positively correlated with reduced perceived stress, suggesting that this reduction in density may have beneficial effects with respect to stress response (Hölzel et al., 2019). However, this study did not have a control group to which the meditation group could be compared, thus it is difficult to make conclusions. A later study was conducted on the same 8-week MBSR program using the exact structural index and had a control group of healthy participants to compare against the meditation group. It was found that individuals in the meditation group showed more increases in density for the left hippocampus, posterior cingulate cortex, left temporoparietal junction, and cerebellum than those of the control group (Hölzel et al., 2011). It is interesting to see that the hippocampus is again found to improve, although this study found the left hippocampus to show an increase in density, rather than the right hippocampus. Another cross-sectional study of experienced meditators (a mixture of techniques were involved) and controls found that experienced meditators had greater density in the bilateral hippocampus, consistent with the trend found in a different volumetric study (Luders et al., 2009) that included some of the same participants from the present study (Luders et al., 2013). In sum, the hippocampus seems to be affected by short-term and long-term meditation experience.

Gyrification

Only one cross-sectional study has examined gyrification in meditation studies. Similar to the previous study (Luders et al., 2013) using gray matter density as the target index, this study used the same group of participants (experienced meditators vs controls) and found greater gyrification in experienced meditators for the bilateral anterior insula, left precentral gyrus, right fusiform gyrus, and right cuneus than the control group (Luders, Kurth et al., 2012; Luders, Phillips et al., 2012), suggesting potential increases in cortical surface area for these brain regions among experienced meditators. However, this study is warranted further replication using a randomized controlled trial (RCT) design.

Brief summary of structural studies of gray matter

Despite each structural index tapping into somewhat different underlying neuronal and cellular mechanisms, there are brain regions that are consistently being found across all indexes to be greater in experienced meditators than in matched healthy controls. Notably, the bilateral hippocampus, bilateral insula, and left inferior temporal gyrus are the most affected structures. We will discuss their functional relevance to behavior with more details in later sections of this chapter. It is also interesting to note that a small number of short-term meditation studies also found structural changes, which may suggest that short-term and long-term experiences could be beneficial to gray matter plasticity, but more rigorous research is needed to validate these findings as having waitlist controls is not the ideal research design when we seek to examine and compare before versus postintervention changes (Tang, Hölzel et al., 2015).

Structural connectivity studies of meditation

As we briefly discussed, white-matter tracts and fiber bundles enable communications across different regions in the brain and hence can be seen as the physical basis of brain structural connectivity. Interestingly, not a lot of scholarly attention has been devoted to examining white-matter plasticity in relation to meditation experiences, even though prior reports of other physical and mental training (e.g., juggling) have suggested white matter to be altered by such experiences. In a series of RCT studies conducted by Tang et al. (2010) and Tang, Lu, Fan, Yang, & Posner, 2012 enhanced fractional anisotropy was found in novices who underwent 1-month (twenty 30-min sessions) of integrative body—mind training (IBMT), a type of meditation training that involves a set of different mindfulness techniques to foster effortless attention and awareness of present-moment experiences. In particular, compared to participants from an active control group that underwent relaxation training (involves the relaxation of each muscle group), participants in the meditation group showed greater fractional anisotropy in the left anterior corona radiate, bilateral superior corona radiate, genu of corpus callosum, body of corpus callosum, and left superior longitudinal fasciculus. The white-matter tracts are situated mostly along the midline regions of the brain, passing through both hemispheres as well as the ACC, which has been already implicated in structural MRI studies of experienced meditators. A separate cross-sectional study of white-matter connectivity in experienced meditators (a mixture of meditation techniques were involved) found enhanced fractional anisotropy across 18 white-matter tracts throughout the brain when compared to healthy matched controls, and the most notable group differences included the superior longitudinal fasciculus (Luders, Clark, Narr, & Toga, 2011). Another set of analyses by Luders et al., identified increased fractional anisotropy in the corpus callosum for experienced meditators Luders, Phillips et al. (2012). Interestingly, both the corpus callosum and superior longitudinal fasciculus are white-matter tracts that were found previously by Tang et al., (2010, 2012) for novices who received short-term mindfulness meditation. These results suggest that those two white-matter tracts are likely to be the most sensitive white-matter structures to meditation effects, regardless of the kind of meditation program or technique.

How does meditation induce brain white-matter changes? Our RCTs have suggested that two factors may contribute to brain white matter changes following short-term IBMT, namely (1) axonal density and (2) myelination. About 5 h IBMT within a 2-week period induces axonal density, but 10 h IBMT within a 4-week period leads to both myelination and axonal density (Tang et al., 2012). Our recent work integrated animal and human models to explore the mechanisms of white-matter plasticity following meditation. Specifically, improved brain plasticity through meditation depends on eliciting deeper meditative states with lower frequency and greater amplitude EEG during and following training. Both brain alpha and theta power have been suggested as key biological signatures that potentially facilitate a successful meditative state. However, the exact role of how alpha and theta waves contribute to the initiation and maintenance of a meditative state remains elusive. In accordance with previous studies in humans, we propose that frontal midline theta activity indexes the control needed to maintain the meditation state; whereas alpha activity is related to the preparation needed to achieve the meditative state. Without enough mental preparation, one often struggles with and has difficulty achieving a meditative state, which thus interrupt the processes of brain plasticity. Based on animal studies, the underlying cellular mechanisms involve increased oligodendrocyte proliferation which results in lowered g ratio and an increase in myelination. These human and animal studies provide evidence supporting the hypothesis that meditation induces white-matter changes through increasing frontal midline theta activity and shed light on how to effectively enhance brain plasticity through meditation (Tang, Tang, Rothbart, & Posner, 2019).

Functional relevance of structural changes

We extensively discussed structural studies of brain gray and white matter in relation to meditation practices and interventions. The most important question is how these structural changes are related to behavioral changes, or whether these structural changes have clear benefits for psychological well-being and cognitive function. For gray matter studies, the ACC, insula, inferior temporal gyrus, and hippocampus are often found in long-term and short-term meditators to be greater in terms of structural indexes such as volume, concentration, and cortical thickness. Based on prior work investigating the functionality of these regions, it is not surprising that meditation-related changes were detected. The ACC's role in attentional control and self-regulation has been repeatedly suggested in the literature (Fox et al., 2014; Tang, Hölzel et al., 2015). For meditation practices, regardless of which tradition or technique, attentional control is heavily engaged as part of the practice and individuals need to constantly regulate their emotions, thoughts, and attention tactfully throughout the process. Additionally, the ACC is also related to conflict monitoring, which may suggest that through repeated meditation practices, individuals are more effective in conflict monitoring such that they are able to promptly notice their mind wandering (in conflict with the practice goal) and directing their attention back to the present focus, consistent with our findings that short-term IBMT improves conflict resolution and emotion regulation accompanied by ACC activity and connectivity (Tang et al., 2007, 2010, 2012). Insula is another highly relevant brain region for meditation practices as it has been consistently shown to support interoceptive awareness such as bodily internal and visceral states as well as metacognitive awareness (Fleming & Dolan, 2012; Fox et al., 2014; Tang, Hölzel et al., 2015; Tang, Lu et al., 2015). These two types of awareness are often exercised in meditation practices, since during meditation individuals are taught to notice their bodily sensations and states, while also becoming aware of their present-moment experiences. Therefore enhanced gray matter in the insula could suggest an improved ability in self-awareness. The inferior temporal gyrus is implicated in high-level visual processing, which may not seem to be intuitive when we consider its functional role in relation to meditation practices. However, there have been some theories suggesting that meditation often involves visual imagery in certain traditions and practices, which means that enhanced gray matter may be related to such repeated practices (Fox et al., 2014). It is unclear whether enhanced gray matter in the inferior temporal gyrus is associated with any obvious behavioral benefits. Finally, the hippocampus has been associated with different processes such as learning, memory, and stress response. Structural abnormalities within the hippocampus are also related to psychiatric disorders, notably depression and anxiety (Fox et al., 2014). It has been suggested that increased gray matter in the hippocampus may serve as a protective factor against stress and may also be an indicator of improved clinical symptoms related to depression and anxiety (Chiesa & Serretti, 2010). Improved psychological health and clinical symptoms are often observed following meditation intervention, thus structural increases in the hippocampus may be related to these psychological benefits.

For white matter, most changes are found in the corpus callosum and superior longitudinal fasciculus, as well as the anterior and superior corona radiate. With regard to the corpus callosum, fractional anisotropy improvement are mostly in the anterior portion, meaning that these altered tracts are usually connected to the prefrontal cortex that encompass regions such as the lateral prefrontal cortex and ACC. These prefrontal regions critical for cognitive control processes have been found to be involved in meditation practices in functional MRI studies, potentially suggesting better connectivity due to meditation experiences. Furthermore, the corpus callosum connects both hemispheres and increases in connectivity could also suggest more efficient interhemispheric communication. The superior longitudinal fasciculus is a major pathway connecting the frontal and posterior portions of the brain and has three subcomponents branching out to temporal and parietal regions of the brain. It appears to be related to higher-order spatial processing and attention (Fox et al., 2014), though its functional role does not seem to correspond to the commonly found improvement in cognitive function. The exact behavioral implications of these white matter changes would need more thorough investigation as we currently have little knowledge about how they may translate into behavior improvement in health and cognition (Tang, Hölzel et al., 2015; Tang, Lu et al., 2015).

Tentative conclusions can be made with regard to the observed structural changes in brain gray and white matter, yet we still lack direct empirical evidence that link structural changes to observable behavior. A lot of the existing interpretations were made based on functional neuroimaging studies of these brain regions and it remains unknown whether more is indeed better with respect to psychological health and cognitive function. Nonetheless these studies do seem to suggest that long-term and short-term meditation practices could influence brain structure to some extent and that more practice time seems to be related to greater changes in brain gray and white matter (Fox et al., 2014; Tang, 2017; Tang, Hölzel et al., 2015; Tang, Lu et al., 2015).

Functional magnetic resonance imaging studies of meditation

Reviewing functional MRI studies of meditation could be a daunting endeavor if we did not discuss these studies within a theoretical framework. For structural MRI studies of meditation, we have already touched upon some of the critical processes involved in meditation: attention control and self-awareness. In a review on the neuroscience of meditation, Tang, Hölzel, et al. (2015) proposed a conceptual framework of meditation that includes attention control, emotion regulation, and self-awareness as core processes involved in meditation. Emotion regulation is imperative for meditation practices since regulating emotion fluctuations during the observation of internal thoughts and sensations is necessary for maintaining the focus on present-moment experiences with an accepting and nonjudgmental attitude. Taking into account existing behavioral and neuroscientific literature, this framework serves as an appropriate starting point for our discussion of functional MRI studies of meditation. It should be noted that our review of neuroscientific literature is not intended to be exhaustive, but rather a discussion of exemplary studies involving major meditation techniques and intervention in relation to the conceptual framework.

Attention control and meditation

Many meditation traditions and techniques involve training attention control to effectively focus on the object of observation during the practice. It is one of the first regulatory process that is heavily emphasized very early on in learning meditation practice (Lutz, Slagter, Dunne, & Davidson, 2008; Tang, Hölzel et al., 2015; Tang, Lu et al., 2015; Tang & Posner, 2009). Throughout the practice, attention control is engaged to maintain the focus on present-moment experiences and previous studies have reported that behaviorally, people showed improved attention control following short-term meditation intervention (Tang, Hölzel et al., 2015; Tang, Lu et al., 2015; Tang et al., 2007). According to established theories of attention, attention can be divided into three components: (1) alerting, which is the readiness in preparation for an upcoming stimulus, involving vigilance as part of the process; (2) orienting, which is the ability to direct and select relevant information from multiple incoming stimuli; and (3) conflict monitoring, which is the ability to monitor and resolve conflict arising from processing the information and is also referred to as executive attention (Petersen & Posner, 2012). There are also other types of attention such as sustained attention, which can be characterized as combining both alerting and orienting attention that facilitates the maintenance of vigilance throughout the tasks. With regard to neural correlates of these three components of attention, alerting involves the brain's noradrenaline system, which originates in the locus coeruleus, whereas orienting attention involves frontal and parietal areas of the brain, including the frontal eye fields and inferior and superior parietal lobes. Executive attention (or conflict monitoring and resolution) involves the ACC, anterior insula, and basal ganglia (Petersen & Posner, 2012; Raz & Buhle, 2006).

Behavioral studies of different meditation techniques and interventions have yielded mixed findings with respect to the improvement of all three components of attention, in that some intervention showed improvement in alerting, while others showed improvement in orienting or conflict monitoring (or executive attention). Specifically, better performance in conflict monitoring has been shown in both novices who received short-term meditation intervention and experienced meditators in cross-sectional studies (van den Hurk, Giommi, Gielen, Speckens, & Barendregt, 2010). However, alerting usually does not show significant improvement after short-term meditation intervention, but long-term meditation experiences have been shown to benefit this aspect of attention (Tang, Hölzel et al., 2015; Tang, Lu et al., 2015). For orienting, improvement has also been mostly found for longer periods of intervention (3-month of Shamatha meditation), whereas 2-month of MBSR did not yield similar effects. Based on a systematic review of attention studies conducted among meditation practitioners, the conclusion appears to be that in the early phases of meditation, improvements are likely to be found in conflict monitoring and orienting, whereas later phases might be associated with more improvement in alerting (Chiesa, Calati, & Serretti, 2011). However, 1-week (five 30-min sessions) of IBMT can lead to improved conflict monitoring (or executive attention), but 4-week (twenty 30-min sessions) of IBMT improves both conflict monitoring and alerting (Tang et al., 2007, 2009). Interestingly, the brain regions involved in conflict monitoring (or executive attention) have consistently been shown across different meditation traditions and techniques. For instance, a fairly recent meta-analysis of functional MRI studies of meditation has implicated the ACC and insula as brain regions critical for most meditation traditions and techniques (Fox et al., 2016). Accompanied by the improvements in executive and alerting attention, later longitudinal studies showed that short-term IBMT during resting state induces greater regional blood flow for ACC and insula than those in the relaxation training group (Tang et al., 2009; Tang, Lu et al., 2015). These findings may indicate that meditation techniques seem to play an important role in modifying attention networks. If the meditation directly targets the ACC and insula, it may first improve executive attention and then alerting attention. This proposal has received support from other meditation studies. For instance, following an 8-week MBSR, a cross-sectional study in novices detected orienting attention improvement using the same attention network task as in the IBMT studies (Jha, Krompinger, & Baime, 2007). One possibility may be that novices mainly used the body scan based concentration meditation technique (e.g., requiring participants to focus on, and move attention from, one body part to another) and led to orienting change; in contrast, in the same study the participants in a 1-month retreat group demonstrated altered performance on the alerting component in comparison with the waitlist control and 8-week MBSR participants. The three groups did not differ in conflict monitoring performance after 1 or 2-month periods of meditation (Jha et al., 2007). Similarly, other cross-sectional studies on long-term experienced meditators who practice one specific meditation technique (focused attention) have shown that the ACC and insula exhibit greater activation compared to healthy matched control groups (Fox et al., 2016). Focused attention involves directing attention to an object of focus (which is typically breathing or a specific body part) and maintaining this focus to minimize mind-wandering. These findings together provide converging evidence that the critical engagement of attention control in meditation are related to the ACC and insula, which also showed enhanced activation after repeated practices, indicating that they may be the underlying mechanisms of behavioral improvement in attention control, especially for executive attention.

Other areas within the prefrontal cortex relevant for attention control are also worth noting. The dorsolateral prefrontal cortex has been shown to have greater activation across different techniques of meditation (including focused attention) (Fox et al., 2016). For instance, for short-term meditation intervention that taught various meditation techniques, greater activation in dorsolateral prefrontal cortex was detected in the meditation group than the active control group during a cognitive task involving executive processing (Allen et al., 2012). Likewise, for patients with generalized anxiety disorder, greater activation in the ventrolateral prefrontal cortex was found in patients who underwent 2-month MBSR than those in an active control group that received stress management education (Hölzel et al., 2013). Additionally, the increased activation in ventrolateral prefrontal cortex was correlated with improvement in anxiety symptoms, suggesting that such increased activation in prefrontal cortex may underlie improvement in psychological well-being. Despite the evidence that brain regions relevant for attention control and regulation show functional and structural changes following short-term and long-term meditation practice, we have yet to determine whether these brain changes are causally related to the improved attentional performance or whether these brain changes are the precise mechanisms underlying behavioral improvement. For example, when are the ACC or dorsolateral prefrontal cortex involved in meditation? Some evidence suggested mental effort during meditation as relating to these brain areas respectively——ACC is dominant when using less effort (or effortless concentration) to meditate whereas the dorsolateral prefrontal cortex engages during effortful attention to meditate, which is consistent with previous findings (Bush, Luu, & Posner, 2000; Tang, 2017; Tang, Hölzel et al., 2015; Tang, Lu et al., 2015; Tang et al., 2012). More rigorous longitudinal neuroimaging studies that employ an active control group as comparison, as well as direct measures of attentional performance are needed to validate these findings.

Emotion regulation and meditation

Emotion regulation refers to the ability to effectively exert control over one's emotions through a wide range of strategies to influence which emotions one has, experiences, or expresses (Gross, 2001). Improvement in psychological well-being and health are the most widely observed benefits as a result of meditation practice and intervention. For instance, self-report studies of psychological well-being have generally found a positive affect to increase and negative affect to decrease among both long-term and short-term meditation practitioners who engaged in various traditions and techniques (Gu, Strauss, Bond, & Cavanagh, 2015). Additionally, physiological measures have demonstrated lower stress reactivity for meditators than for active controls after both groups were introduced to stressful scenarios (Tang et al., 2007). Theoretically, enhanced emotion regulation capacity is thought to underlie these psychological and physiological benefits associated with meditation, as negative affect, stress reactivity, and symptoms of psychological disorders are primarily due to deficits in emotion regulation (Tang, Hölzel et al., 2015; Tang, Lu et al., 2015). According to well-established theoretical models of emotion regulation, attention (allocating resources to regulate emotion), cognition (alter the appraisal of emotion), and inhibition (suppress emotion) are often involved in order to effectively regulate emotions (Tang, Hölzel et al., 2015; Tang, Lu et al., 2015). For meditation, most techniques would involve these three processes to some extent. For example, observing present-moment experiences during meditation would require attending to any emotion, thought, or sensation that arises at each moment. Such attentive observation of emotional experiences is achieved with a nonemotionally reactive and nonjudgmental attitude, which requires the engagement of cognitive reappraisal to construct emotionally-aroused experiences into emotion-detached events (Hölzel et al., 2011). One study also suggests that meditation not only elicits cognitive reappraisal as one of the emotion-regulation strategies during the actual practice, but also increases self-report positive reappraisal postintervention (Garland, Gaylord, & Fredrickson, 2011).

Neural correlates of emotion regulation have mainly been found in the limbic system (mostly amygdala) and frontal areas of the brain. The hypothesis for emotion regulation in meditation seems to be that attention and cognitive control are strengthened through prefrontal regions of the brain, which exert a top-down regulation of the limbic system to effectively influence emotion processing (Hölzel et al., 2011; Tang, Hölzel et al., 2015; Tang, Lu et al., 2015). Not surprisingly, functional neuroimaging studies of meditation have found these regions to be affected by meditation experiences. Most studies have asked participants to complete emotion tasks in order to identify differences or changes in brain activation patterns and have generally shown that meditation practices are related to decreases in the activation of the amygdala in response to emotional stimuli (Desbordes et al., 2012; Goldin & Gross, 2010; Lutz et al., 2013), suggesting a lower emotional reactivity or arousal for meditators than for controls. In addition to the limbic system, prefrontal regions such as the dorsolateral prefrontal cortex have shown enhanced activation during emotion processing for individuals who underwent short-term meditation intervention that involved a mixture of different techniques (Allen et al., 2012). Enhanced activation in dorsomedial and dorsolateral prefrontal cortex was observed when participants were told to anticipate negative images while they engaged in a state of mindfulness (Lutz et al., 2014). Of note, these participants were novices who received short-term meditation intervention. However, for experienced meditators, studies on pain suggested that they tend to have lower activation in these prefrontal regions relevant for control and regulatory processes, which may imply that they have a lower need for effortful cognitive control of emotion, instead, they may have a better capacity of accepting these affective experiences (Tang, Hölzel et al., 2015; Tang, Lu et al., 2015). Pain can often evoke negative affect and emotional disturbances, thus can be thought of as a negative emotional and sensational experience. Effectively regulating pain sensitivity and response would undoubtedly involve emotion regulation to some extent. In one cross-sectional study, experienced meditators who practice Zen meditation had decreased activation in the dorsolateral and ventrolateral prefrontal cortex when experiencing thermal pain compared with healthy controls

(Grant, Courtemanche, & Rainville, 2011) and that lower-pain sensitivity was positively correlated with decreased activation in these brain regions. Another study also reported that long-term meditators showed decreased activation in the lateral prefrontal cortex and increased activation in the insula during pain stimulation, which were both correlated with lower pain-related unpleasantness (Gard et al., 2011). These findings support the notion that novice meditators may require more engagement from the prefrontal cortex for effortful and deliberate control of emotions, while experienced meditators may engage in effortless control of emotion and have a more accepting attitude toward their experiences, thus exhibiting decreased activation in prefrontal regions, but potentially more activation in regions (e.g., insula, striatum) associated with self-awareness and reward (Hölzel et al., 2011; Tang, Hölzel et al., 2015; Tang, Lu et al., 2015).

This hypothesis is also corroborated by neuroimaging studies of loving kindness meditation. Loving kindness meditation (or compassion) seeks to foster a sense of sympathy, love, friendliness, warmth, and kindness toward oneself and others, which has been suggested to promote altruistic behaviors (Fox et al., 2016; Lutz, Brefczynski-Lewis, Johnstone, & Davidson, 2008). Intuitively, loving kindness meditation requires the engagement of emotion regulation to enable positive emotion throughout the practice. Interestingly, a meta-analysis of neuroimaging studies on loving kindness meditation have identified the anterior insula to show enhanced activation (Fox et al., 2016), indicating that increased activation in the insula may suggest effective emotion regulation and could also be the mechanism underlying emotion regulation–related benefits in experienced meditators. Based on our discussion thus far, the insula seems to be implicated in both attention control and emotion regulation, suggesting that it could well be an important neural substrate of improved behavioral outcomes in attention and emotion following either long-term or short-term meditation experiences. Studies directly linking behavior with neuroimaging findings remain limited, which warrants more empirical investigations to replicate and validate these results.

Self-Awareness and meditation

Awareness is an active ingredient of meditation practices. Typically, awareness in meditation can be directed to bodily sensations, objects of focus, or thoughts that constantly come and go through our consciousness during practice. In general, most meditation techniques and traditions exercise awareness, whether it is the awareness that is being guided toward oneself, or toward the surrounding around oneself. Considering the nature of meditation, awareness is just as indispensable as attention is to meditation practices, which together establish the foundation of achieving a meditative state. The term self-awareness within meditation can represent both awareness of present-moment experiences, which frequently involve interoceptive awareness concerning the practitioner, as well as awareness of the "self" or, in some sense, can be defined as awareness of self-referential processing (Tang, Hölzel et al., 2015; Tang, Lu et al., 2015). For awareness of present-moment experiences, individuals who practice meditation often report greater awareness of bodily sensations and more perceptual clarity of subtle internal bodily states. They also report more self-awareness in their daily lives, which can be assessed through subscales of trait mindfulness questionnaires. Studies using objective measures have yielded mixed findings. In two of the studies, meditators did not seem to show superior interoceptive awareness in a heartbeat detection task when compared against healthy controls (Khalsa et al., 2008; Nielsen & Kaszniak, 2006). However, experienced meditators do seem to have better coherence between their subjective reports and objective assessment with regard to emotional experience and bodily sensations (Tang, Hölzel et al., 2015; Tang, Lu et al., 2015). For awareness of the self, any processes related to self-representation or the concept of self can be categorized as self-related processes that are in reference to the self. Our own concept of self is rather complicated, which in some sense is purely a subjective experience. Most people would consider thoughts, emotions, personality, and characteristics as part of the self. Our body and mind, and even anything we label as ours, can also be considered as part of the self. There seem to be three types of selves when we think about the concept. The first could be the cognitive self that mainly involves our values, beliefs, and thoughts. There is the second type of self, our body, which involves bodily sensations such as interoception. The third self is the phenomenal-experiential self that can be conceptualized as the self in one's present awareness (Tang, 2017). During meditation, our cognitive self and bodily self are being observed in a detached manner. Our self in the present awareness is being elevated through meta-awareness such that we are not only aware of the present moment, but are also aware of our present awareness. During this process the concept of self and self-representation are amplified and can be reconstructed once individuals notice the different layers of self. According to Buddhist philosophy, our perception of the self would change over the course of meditation practices as we gradually realize that our experiences such as thoughts and emotions are constantly changing, which may lead us to alter our identification with a static self (Tang & Tang, 2013, 2015; Tang, Hölzel et al., 2015; Tang, Lu et al., 2015; Tang, 2017). For most meditation techniques this distinction of different selves comes in later stages, especially among experienced meditators.

Neural correlates of self-awareness, especially for interoceptive awareness, have largely been found in the insula based on prior literature (Tang, Hölzel et al., 2015; Tang, Lu et al., 2015). Not surprisingly, most meditation studies have detected increased activation in the insula across various traditions and techniques, although a few studies have shown direct connection between enhanced insula activation and heightened awareness in behavior

(Fox et al., 2016). In one study of novices receiving short-term MBSR, greater insula activation was found during interoceptive attention to respiratory sensations compared to healthy controls (Farb et al., 2012). One specific meditation technique, open monitoring, largely involves self-awareness in its practice. Open monitoring does not involve a specific object of focus, instead, the object of focus is anything that arises at the present moment. Hence the attention and awareness during open-monitoring practice can be considered as broad and unconstrained, yet the key is to be nonjudgmental of the experiences in an equanimous way. Interestingly, open monitoring not only involves the insula, but also recruits regions from the prefrontal cortex such as the dorsolateral portion as well as the dorsal ACC (Tang, Lu et al., 2015; Fox et al., 2016). For instance, brief IBMT practice used the same open-monitoring technique and results suggested self-awareness is supported by the insula and ACC (Tang, Lu et al., 2015). These regions together could support the necessary attention resources and awareness for present-moment experiences as they are responsible for voluntary regulation of thoughts and action.

The neural basis of self-referential processing (e.g., mind-wandering) can be largely found within the default mode network, which involves anterior—posterior midline structures of the brain including the medial prefrontal cortex, posterior cingulate cortex, anterior precuneus, and inferior parietal lobule (Greicius, Krasnow, Reiss, & Menon, 2003; Raichle et al., 2001). Based on functional neuroimaging studies of the brain at rest, this network is highly active when there is no ongoing cognitive-task demand. When individuals engage in demanding tasks, deactivation of this network is often observed. A meta-analysis of the neural substrates of self and its related processes have indicated that the regions in the default mode network are primarily responsible for the concept of self and self-representation (Northoff et al., 2006). Therefore any potential alterations of the "self" (i.e., detaching from cognitive and bodily selves) following meditation experiences should manifest in the brain, if any, within this default mode network. Indeed, in one study of experienced meditators practicing different meditation techniques the default mode network regions showed deactivation, but greater functional connectivity was found among the posterior cingulate cortex and dorsal ACC and the dorsolateral prefrontal cortex than health controls at baseline (Brewer et al., 2011). The deactivation of the posterior cingulate cortex for experienced meditators during meditation practices can be interpreted as less mind-wandering (e.g., less engagement of the cognitive self) or goal-irrelevant activity as practicing meditation is like completing a "task" to some extent. Furthermore, the strong coupling between region of the default mode network and regions relevant for attention and cognitive control suggest that reduced mind-wandering may be achieved through enhanced attention control and that this coupling may be the mechanism underlying this reduction in self-referential processing.

Brain networks and meditation

Based on our review of structural and functional neuroimaging studies of meditation, it is evident that multiple regions have been consistently found across novices who receive short-term meditation intervention as well as in experienced meditators who have years of practice experience. A quick survey of the neuroimaging literature on meditation would reveal that meditation-related changes are primarily found in brain regions that fall within three established functional networks associated with attentional and cognitive control and interoceptive awareness: frontoparietal network, default mode network, and cingulo-opercular/salience network. Although most inferences on the underlying neural mechanism of meditation have focused on the role of isolated brain regions in enhancing behavioral outcomes, it is imperative to consider the networks to which these regions belong or their interactions with other regions in those networks that may concurrently support the observed psychological and cognitive benefits. For instance, the dorsolateral prefrontal cortex in the frontoparietal network is associated with enhanced attention and cognitive control following meditation, yet the role of the frontoparietal network as a key network supporting cognitive control processes is not well-explored or mentioned. Another example is the dorsal/ventral ACC and anterior insula, which in separate studies exhibited heightened blood flow and were associated with increased capacity for conflict resolution and executive function following meditation. Both regions are key nodes of the cingulo-opercular/salience network, a network that is responsible for switching and monitoring the states of frontoparietal network and default mode network (Tang, Rothbart, & Posner, 2012) and presumably allowing the allocation of attentional resources to support cognitive control. Nevertheless these neuroimaging studies of meditation were not able to draw neural mechanistic conclusions at a network level to bridge the frontoparietal network, default mode network, and cingulo-opercular/salience network together into a coherent framework that explains the cognitive effects of meditation. Our recent large-scale network study revealed that brief IBMT reorganized whole-brain networks related to attention, cognitive and affective processing, awareness and sensory integration, and reward processing (Tang et al., 2017) (see the discussion in Chapter 8: How to measure outcomes and individual differences in meditation). Further research examining meditation effects and outcomes at the network level would be necessary to advance our understanding.

Notably, there has been a shift in focus within cognitive neuroscience from isolated brain regions to large-scale brain networks (consisting of key regions that are temporally correlated with one another), which has greatly contributed to our understanding of the mechanisms underlying psychological and cognitive processes. In light of this new perspective it is highly likely that various mental states such as the meditative state or mental state-related changes such as those induced by meditation, may manifest themselves more explicitly through brain networks (Tang, Hölzel et al., 2015; Tang, Lu et al., 2015). Currently not a lot of scientific investigations have examined the neural correlates and mechanisms of meditation at the level of brain networks. We will discuss this network perspective in later chapters of this book along with useful methodologies.

Individual differences in brain functional regions, networks, and meditation

Everyone's brain is different. Although it is less known with respect to how individual differences in brain structures and functional activity and connectivity may translate into differences in behavior, we would like to briefly discuss one functional network in relation to its putative implications for meditation practices and outcomes. We take the frontoparietal network as an example since it has been demonstrated quite reliably as a crucial network directly supporting cognitive function and various task demands by prior research (Zanto & Gazzaley, 2013). Furthermore, our discussion has also repeatedly described regions within the frontoparietal network to be actively involved in meditation practices and also being altered by meditation. Therefore the following discussion will use the frontoparietal network as an example to illustrate the role of preexisting individual differences in influencing meditation-related improvement in the brain and behavior.

Previous studies have consistently shown enhanced activation of the dorsolateral prefrontal cortex in the frontoparietal network in novice participants who underwent meditation intervention (Allen et al., 2012; Tomasino & Fabbro, 2016), while experienced meditators exhibited high connectivity between the dorsolateral prefrontal cortex and other brain regions in the default mode network and cingulo-opercular/salience network (Brewer et al., 2011). Interestingly, the lateral prefrontal cortex (encompassing the dorsolateral prefrontal cortex) is a connector hub region in the frontoparietal network that enables across-network communication by flexibly adapting its connectivity with other functional networks in response to specific task demands (Cole, Ito, & Braver, 2015). Moreover, the strength of connectivity of lateral prefrontal cortex with regions in different functional networks is found to contribute to individual variability in fluid intelligence and cognitive control (Cole et al., 2015; Cole, Yarkoni, Repovš, Anticevic, & Braver, 2012), such that high connectivity predicts better performance in tasks involving fluid intelligence and cognitive control. Although these results suggest that meditation may enhance cognitive function through increasing activation and connectivity of lateral prefrontal cortex (Tang, Hölzel et al., 2015; Tang, Lu et al., 2015), preexisting individual variabilities in these neural indexes may impose a physical constraint on the extent of change in brain and cognitive function. Specifically, individuals with inherently high levels of lateral prefrontal cortex connectivity and activation may have already been functionally configured to an optimal condition in the brain, such that there is simply no room for further improvement. The same is likely true for other brain regions exhibiting considerable interindividual variabilities, such as the posterior cingulate cortex in the default mode network and the dorsal ACC in the cingulo-opercular/salience network, whose connectivity and patterns of activation have found to be altered by meditation practices (Tang, 2017; Tang, Hölzel et al., 2015; Tang, Lu et al., 2015). Both relevant to processes of cognitive control (Leech & Sharp, 2013; Shenhav, Botvinick, & Cohen, 2013). Therefore studies utilizing changes in brain connectivity and patterns of activation as indexes of meditation effects or making inferences about behavioral and cognitive improvement based on these brain changes would need to be mindful of inherent individual differences presented in these measures, rather than simply averaging data with substantive individual variability to draw conclusions.

In addition to individual differences in specific neural substrates of cognitive processes at the region-wise level, there are also individual-specific features in higher-level brain network organization that may modulate the extent of training-related improvement in brain and behavior. An impressive body of literature regarding individual differences in brain functional networks has identified not only intrinsic network architecture shared by the population, meaning that commonly defined networks are present across individuals, but also stable individual-specific network characteristics that vary depending on task demands (Finn et al., 2015; Gordon, Laumann, Adeyemo, & Petersen, 2017; Gratton et al., 2018). In particular, one study illustrated that task-evoked modulation of functional networks primarily behaves in an individual-specific manner, such that the greatest individual variability can be observed in the "control systems," especially in the fronto-parietal regions relevant for high-level cognitive processes; whereas the "processing systems" (i.e., sensorimotor regions) showed the most similarity across individuals (Gratton et al., 2018). As most meditation studies are concerned with alterations in the "control systems" and their associated cognitive processes during practice or at postintervention, collapsing data across individuals are most likely to eliminate important information regarding individual level training effects on brain functional networks, thereby resulting in inaccurate estimations and conclusions of the magnitude of effects on the brain as well as on behavior if group-averaged effects on the brain are used to draw inferences about the extent of change in behavior. Relatedly, another source of potential confound arises from the fact that individual-specific, task-evoked effects on brain functional networks account for about 20% of variance, compared to only 5% of variance explained by cross-individual task-specific modulations of brain networks (Gratton et al., 2018), which suggests that individual-specific analyses may be more appropriate and sensitive for detecting any meditation-related changes in brain functional networks while completing tasks, as opposed to group averaging approaches that may obscure potentially large effects induced by meditation intervention. Although most meditation studies have yet to employ network-based analyses in examining intervention effects, this discussion provides some cautionary notes for typical meditation studies investigating changes in task-evoked functional connectivity that often overlook the role of individual differences, making them susceptible to small cross-subject task-specific effects that cannot be easily detected when a group-averaging approach is employed. Further, employing an individual differences approach in neuroimaging studies of meditation holds promise for identifying network-level biomarkers that may be more sensitive to effects of meditation and those of other forms of psychological intervention than region-wise brain activation or functional connectivity.

Summary and implications for translational research

Taken together, individual differences are ubiquitous in almost all domains relevant for human functioning, ranging from dispositional traits to patterns of brain activation and functional networks. We discussed critical brain regions and networks relevant for meditation practices and outcomes and also explained how individual differences in the brain may subsequently influence the effectiveness and outcomes of meditation-based interventions. These observed individual differences are not only robust and fairly reliable as described throughout the book, but are critical for scientific investigation and effective application of psychological intervention in preventing and treating problematic behavior in individuals. For instance, translational research would undoubtedly involve identifying neural correlates and markers associated with meditation interventions, which means that attending to individual differences at the brain level is equally as important as considering such differences at the behavioral level. Given that increasing efforts are being devoted to individual variability within brain structures and functions, future translational research may be able to examine individual outcomes of the brain at an individual basis. More research efforts need to be devoted into identifying putative individual differences in the brain relevant for predicting intervention outcomes and understanding how these individual variabilities interact with psychological interventions to produce differential effects.

References

Allen, M., Dietz, M., Blair, K. S., van Beek, M., Rees, G., Vestergaard-Poulsen, P., ... Roepstorff, A. (2012). Cognitive-affective neural plasticity following active-controlled mindfulness intervention. *The Journal of Neuroscience*, *32*(44), 15601–15610.

Brewer, J. A., Worhunsky, P. D., Gray, J. R., Tang, Y. Y., Weber, J., & Kober, H. (2011). Meditation experience is associated with differences in default mode network activity and connectivity. *Proceedings of the National Academy of Sciences of the United States of America*, *108*(50), 20254–20259.

Bullmore, E., & Sporns, O. (2009). Complex brain networks: Graph theoretical analysis of structural and functional systems. *Nature Reviews Neuroscience*, *10*(3), 186–198.

Burle, B., Spieser, L., Roger, C., Casini, L., Hasbroucq, T., & Vidal, F. (2015). Spatial and temporal resolutions of EEG: Is it really black and white? A scalp current density view. *International Journal of Psychophysiology*, *97*(3), 210–220.

Bush, G., Luu, P., & Posner, M. I. (2000). Cognitive and emotional influences in anterior cingulate cortex. *Trends in cognitive sciences*, *4*(6), 215–222.

Cahn, B. R., & Polich, J. (2006). Meditation states and traits: EEG, ERP, and neuroimaging studies. *Psychological Bulletin*, *132*(2), 180–211.

Chiesa, A., Calati, R., & Serretti, A. (2011). Does mindfulness training improve cognitive abilities? A systematic review of neuropsychological findings. *Clinical Psychology Review*, *31*(3), 449–464.

Chiesa, A., & Serretti, A. (2010). A systematic review of neurobiological and clinical features of mindfulness meditations. *Psychological Medicine*, *40*(8), 1239–1252.

Cole, M. W., Ito, T., & Braver, T. S. (2015). Lateral prefrontal cortex contributes to fluid intelligence through multinetwork connectivity. *Brain Connectivity*, *5*(8), 497–504.

Cole, M. W., Yarkoni, T., Repovš, G., Anticevic, A., & Braver, T. S. (2012). Global connectivity of prefrontal cortex predicts cognitive control and intelligence. *Journal of Neuroscience*, *32*(26), 8988–8999.

Desbordes, G., Negi, L. T., Pace, T. W., Wallace, B. A., Raison, C. L., & Schwartz, E. L. (2012). Effects of mindful-attention and compassion meditation training on amygdala response to emotional stimuli in an ordinary, non-meditative state. *Frontiers in Human Neuroscience*, *6*, 292.

Farb, N. A., Segal, Z. V., & Anderson, A. K. (2012). Mindfulness meditation training alters cortical representations of interoceptive attention. *Social Cognitive and Affective Neuroscience*, *8*(1), 15–26.

Finn, E. S., Shen, X., Scheinost, D., Rosenberg, M. D., Huang, J., Chun, M. M., ... Constable, R. T. (2015). Functional connectome fingerprinting: Identifying individuals using patterns of brain connectivity. *Nature Neuroscience*, *18*(11), 1664–1671.

Fleming, S. M., & Dolan, R. J. (2012). The neural basis of metacognitive ability. *Philosophical Transactions of the Royal Society B: Biological Sciences*, *367*(1594), 1338–1349.

Fox, K. C., Dixon, M. L., Nijeboer, S., Girn, M., Floman, J. L., Lifshitz, M., ... Christoff, K. (2016). Functional neuroanatomy of meditation: A review and meta-analysis of 78 functional neuroimaging investigations. *Neuroscience & Biobehavioral Reviews*, *65*, 208–228.

Fox, K. C., Nijeboer, S., Dixon, M. L., Floman, J. L., Ellamil, M., Rumak, S. P., ... Christoff, K. (2014). Is meditation associated with altered brain structure? A systematic review and meta-analysis of morphometric neuroimaging in meditation practitioners. *Neuroscience & Biobehavioral Reviews*, *43*, 48–73.

Fox, M. D., & Raichle, M. E. (2007). Spontaneous fluctuations in brain activity observed with functional magnetic resonance imaging. *Nature Reviews Neuroscience*, *8*(9), 700–711.

Fox, M. D., Snyder, A. Z., Vincent, J. L., Corbetta, M., Van Essen, D. C., & Raichle, M. E. (2005). The human brain is intrinsically organized into dynamic, anticorrelated functional networks. *Proceedings of the National Academy of Sciences of the United States of America*, *102*(27), 9673–9678.

Gard, T., Hölzel, B. K., Sack, A. T., Hempel, H., Lazar, S. W., Vaitl, D., & Ott, U. (2011). Pain attenuation through mindfulness is associated with decreased cognitive control and increased sensory processing in the brain. *Cerebral Cortex, 22*(11), 2692–2702.

Garland, E. L., Gaylord, S. A., & Fredrickson, B. L. (2011). Positive reappraisal mediates the stress-reductive effects of mindfulness: An upward spiral process. *Mindfulness, 2*(1), 59–67.

Goldin, P. R., & Gross, J. J. (2010). Effects of mindfulness-based stress reduction (MBSR) on emotion regulation in social anxiety disorder. *Emotion (Washington, D.C.), 10*(1), 83–91.

Gordon, E. M., Laumann, T. O., Adeyemo, B., & Petersen, S. E. (2017). Individual variability of the system-level organization of the human brain. *Cerebral Cortex, 27*(1), 386–399.

Grant, J. A., Courtemanche, J., Duerden, E. G., Duncan, G. H., & Rainville, P. (2010). Cortical thickness and pain sensitivity in Zen meditators. *Emotion (Washington, D.C.), 10*(1), 43–53.

Grant, J. A., Courtemanche, J., & Rainville, P. (2011). A non-elaborative mental stance and decoupling of executive and pain-related cortices predicts low pain sensitivity in Zen meditators. *Pain, 152*(1), 150–156.

Grant, J. A., Duerden, E. G., Courtemanche, J., Cherkasova, M., Duncan, G. H., & Rainville, P. (2013). Cortical thickness, mental absorption and meditative practice: Possible implications for disorders of attention. *Biological Psychology, 92*(2), 275–281.

Gratton, C., Laumann, T. O., Nielsen, A. N., Greene, D. J., Gordon, E. M., Gilmore, A. W., . . . Dosenbach, N. U. (2018). Functional brain networks are dominated by stable group and individual factors, not cognitive or daily variation. *Neuron, 98*(2), 439–452.

Greicius, M. D., Krasnow, B., Reiss, A. L., & Menon, V. (2003). Functional connectivity in the resting brain: A network analysis of the default mode hypothesis. *Proceedings of the National Academy of Sciences of the United States of America, 100*(1), 253–258.

Gross, J. J. (2001). Emotion regulation in adulthood: Timing is everything. *Current Directions in Psychological Science, 10*(6), 214–219.

Gu, J., Strauss, C., Bond, R., & Cavanagh, K. (2015). How do mindfulness-based cognitive therapy and mindfulness-based stress reduction improve mental health and wellbeing? A systematic review and meta-analysis of mediation studies. *Clinical Psychology Review, 37*, 1–12.

Hölzel, B. K., Carmody, J., Evans, K. C., Hoge, E. A., Dusek, J. A., Morgan, L., . . . Lazar, S. W. (2009). Stress reduction correlates with structural changes in the amygdala. *Social Cognitive and Affective Neuroscience, 5*(1), 11–17.

Hölzel, B. K., Carmody, J., Evans, K. C., Hoge, E. A., Dusek, J. A., Morgan, L., & Lazar, S. W. (2009). Stress reduction correlates with structural changes in the amygdala. *Social cognitive and affective neuroscience, 5*(1), 11–17.

Hölzel, B. K., Carmody, J., Vangel, M., Congleton, C., Yerramsetti, S. M., Gard, T., & Lazar, S. W. (2011). Mindfulness practice leads to increases in regional brain gray matter density. *Psychiatry Research: Neuroimaging, 191*(1), 36–43.

Hölzel, B. K., Hoge, E. A., Greve, D. N., Gard, T., Creswell, J. D., Brown, K. W., . . . Lazar, S. W. (2013). Neural mechanisms of symptom improvements in generalized anxiety disorder following mindfulness training. *NeuroImage: Clinical, 2*, 448–458.

Hölzel, B. K., Ott, U., Gard, T., Hempel, H., Weygandt, M., Morgen, K., & Vaitl, D. (2007). Investigation of mindfulness meditation practitioners with voxel-based morphometry. *Social cognitive and affective neuroscience, 3*(1), 55–61.

Jha, A. P., Krompinger, J., & Baime, M. J. (2007). Mindfulness training modifies subsystems of attention. *Cognitive, Affective and Behavioral Neuroscience, 7*(2), 109–119.

Jones, D. K., Knösche, T. R., & Turner, R. (2013). White matter integrity, fiber count, and other fallacies: The do's and don'ts of diffusion MRI. *NeuroImage, 73*, 239–254.

Khalsa, S. S., Rudrauf, D., Damasio, A. R., Davidson, R. J., Lutz, A., & Tranel, D. (2008). Interoceptive awareness in experienced meditators. *Psychophysiology, 45*(4), 671−677.

Krakauer, J. W., Ghazanfar, A. A., Gomez-Marin, A., MacIver, M. A., & Poeppel, D. (2017). Neuroscience needs behavior: Correcting a reductionist bias. *Neuron, 93*(3), 480−490.

Lazar, S. W., Kerr, C. E., Wasserman, R. H., Gray, J. R., Greve, D. N., Treadway, M. T., & Rauch, S. L. (2005). Meditation experience is associated with increased cortical thickness. *Neuroreport, 16*(17), 1893.

Leech, R., & Sharp, D. J. (2013). The role of the posterior cingulate cortex in cognition and disease. *Brain, 137*(1), 12−32.

Leung, M. K., Chan, C. C., Yin, J., Lee, C. F., So, K. F., & Lee, T. M. (2012). Increased gray matter volume in the right angular and posterior parahippocampal gyri in loving-kindness meditators. *Social Cognitive and Affective Neuroscience, 8*(1), 34−39.

Lomas, T., Ivtzan, I., & Fu, C. H. (2015). A systematic review of the neurophysiology of mindfulness on EEG oscillations. *Neuroscience & Biobehavioral Reviews, 57*, 401−410.

Luders, E., Clark, K., Narr, K. L., & Toga, A. W. (2011). Enhanced brain connectivity in long-term meditation practitioners. *NeuroImage, 57*(4), 1308−1316.

Luders, E., Kurth, F., Mayer, E. A., Toga, A. W., Narr, K. L., & Gaser, C. (2012). The unique brain anatomy of meditation practitioners: Alterations in cortical gyrification. *Frontiers in Human Neuroscience, 6*, 34.

Luders, E., Phillips, O. R., Clark, K., Kurth, F., Toga, A. W., & Narr, K. L. (2012). Bridging the hemispheres in meditation: Thicker callosal regions and enhanced fractional anisotropy (FA) in long-term practitioners. *NeuroImage, 61*(1), 181−187.

Luders, E., Thompson, P. M., Kurth, F., Hong, J. Y., Phillips, O. R., Wang, Y., ... Toga, A. W. (2013). Global and regional alterations of hippocampal anatomy in long-term meditation practitioners. *Human Brain Mapping, 34*(12), 3369−3375.

Luders, E., Toga, A. W., Lepore, N., & Gaser, C. (2009). The underlying anatomical correlates of long-term meditation: Larger hippocampal and frontal volumes of gray matter. *NeuroImage, 45*(3), 672−678.

Lutz, A., Brefczynski-Lewis, J., Johnstone, T., & Davidson, R. J. (2008). Regulation of the neural circuitry of emotion by compassion meditation: Effects of meditative expertise. *PLoS ONE, 3*(3), e1897.

Lutz, J., Herwig, U., Opialla, S., Hittmeyer, A., Jäncke, L., Rufer, M., & Brühl, A. B. (2013). Mindfulness and emotion regulation—an fMRI study. *Social cognitive and affective neuroscience, 9*(6), 776−785.

Lutz, A., Slagter, H. A., Dunne, J. D., & Davidson, R. J. (2008). Attention regulation and monitoring in meditation. *Trends in Cognitive Sciences, 12*(4), 163−169.

Mechelli, A., Price, C. J., Friston, K. J., & Ashburner, J. (2005). Voxel-based morphometry of the human brain: Methods and applications. *Current Medical Imaging Reviews, 1*(2), 105−113.

Nielsen, L., & Kaszniak, A. W. (2006). Awareness of subtle emotional feelings: A comparison of long-term meditators and nonmeditators. *Emotion (Washington, D.C.), 6*(3), 392.

Northoff, G., Heinzel, A., De Greck, M., Bermpohl, F., Dobrowolny, H., & Panksepp, J. (2006). Self-referential processing in our brain—a meta-analysis of imaging studies on the self. *NeuroImage, 31*(1), 440−457.

Pagnoni, G., & Cekic, M. (2007). Age effects on gray matter volume and attentional performance in Zen meditation. *Neurobiology of Aging, 28*(10), 1623−1627.

Patriat, R., Molloy, E. K., Meier, T. B., Kirk, G. R., Nair, V. A., Meyerand, M. E., ... Birn, R. M. (2013). The effect of resting condition on resting-state fMRI reliability and consistency: A comparison between resting with eyes open, closed, and fixated. *NeuroImage, 78*, 463–473.

Petersen, S. E., & Posner, M. I. (2012). The attention system of the human brain: 20 years after. *Annual Review of Neuroscience, 35*, 73–89.

Power, J. D., Cohen, A. L., Nelson, S. M., Wig, G. S., Barnes, K. A., Church, J. A., ... Petersen, S. E. (2011). Functional network organization of the human brain. *Neuron, 72*(4), 665–678.

Raichle, M. E., MacLeod, A. M., Snyder, A. Z., Powers, W. J., Gusnard, D. A., & Shulman, G. L. (2001). A default mode of brain function. *Proceedings of the National Academy of Sciences of the United States of America, 98*(2), 676–682.

Raz, A., & Buhle, J. (2006). Typologies of attentional networks. *Nature Reviews Neuroscience, 7*(5), 367.

Schaefer, A., Kong, R., Gordon, E. M., Laumann, T. O., Zuo, X. N., Holmes, A. J., ... Yeo, B. T. (2017). Local-global parcellation of the human cerebral cortex from intrinsic functional connectivity MRI. *Cerebral Cortex, 28*(9), 3095–3114.

Shenhav, A., Botvinick, M. M., & Cohen, J. D. (2013). The expected value of control: An integrative theory of anterior cingulate cortex function. *Neuron, 79*(2), 217–240.

Tang, Y. Y. (2017). *The neuroscience of mindfulness meditation: How the body and mind work together to change our behaviour.* Springer.

Tang, Y. Y., Lu, Q., Fan, M., Yang, Y., & Posner, M. I. (2012). Mechanisms of white matter changes induced by meditation. *Proceedings of the National Academy of Sciences of the United States of America, 109*(26), 10570–10574.

Tang, Y. Y., Lu, Q., Geng, X., Stein, E. A., Yang, Y., & Posner, M. I. (2010). Short-term meditation induces white matter changes in the anterior cingulate. *Proceedings of the National Academy of Sciences of the United States of America, 107*(35), 15649–15652.

Tang, Y. Y., Hölzel, B. K., & Posner, M. I. (2015). The neuroscience of mindfulness meditation. *Nature Reviews Neuroscience, 16*(4), 213–225.

Tang, Y. Y., Ma, Y., Fan, Y., Feng, H., Wang, J., Feng, S., ... Zhang, Y. (2009). Central and autonomic nervous system interaction is altered by short-term meditation. *Proceedings of the National Academy of Sciences of the United States of America, 106*(22), 8865–8870.

Tang, Y. Y., Ma, Y., Wang, J., Fan, Y., Feng, S., Lu, Q., ... Posner, M. I. (2007). Short-term meditation training improves attention and self-regulation. *Proceedings of the National Academy of Sciences of the United States of America, 104*(43), 17152–17156.

Tang, Y. Y., & Posner, M. I. (2009). Attention training and attention state training. *Trends in Cognitive Sciences, 13*(5), 222–227.

Tang, Y. Y., & Tang, R. (2013). Ventral-subgenual anterior cingulate cortex and self-transcendence. *Frontiers in Psychology, 4*, 1000.

Tang, Y. Y., & Tang, R. (2015). Rethinking future directions of the mindfulness field. *Psychological Inquiry, 26*(4), 368–372.

Tang, Y. Y., Lu, Q., Feng, H., Tang, R., & Posner, M. I. (2015). Short-term meditation increases blood flow in anterior cingulate cortex and insula. *Frontiers in Psychology, 6*, 212.

Tang, Y. Y., Rothbart, M. K., & Posner, M. I. (2012). Neural correlates of establishing, maintaining, and switching brain states. *Trends in cognitive sciences, 16*(6), 330–337.

Tang, Y. Y., Tang, R., Rothbart, M. K., & Posner, M. I. (2019). Frontal theta activity and white matter plasticity following mindfulness meditation. *Current Opinion in Psychology, 28*, 294–297.

Tang, Y. Y., Tang, Y., Tang, R., & Lewis-Peacock, J. A. (2017). Brief mental training reorganizes large-scale brain networks. *Frontiers in Systems Neuroscience, 11*, 6.

Tomasino, B., & Fabbro, F. (2016). Increases in the right dorsolateral prefrontal cortex and decreases the rostral prefrontal cortex activation after-8 weeks of focused attention based mindfulness meditation. *Brain and Cognition, 102*, 46−54.

van den Hurk, P. A., Giommi, F., Gielen, S. C., Speckens, A. E., & Barendregt, H. P. (2010). Greater efficiency in attentional processing related to mindfulness meditation. *Quarterly Journal of Experimental Psychology, 63*(6), 1168−1180.

Wager, T. D. (2006). Do we need to study the brain to understand the mind? *APS Observer, 19* (9), 25−27.

Zanto, T. P., & Gazzaley, A. (2013). Fronto-parietal network: Flexible hub of cognitive control. *Trends in Cognitive Sciences, 17*(12), 602−603.

Zuo, X. N., & Xing, X. X. (2014). Test-retest reliabilities of resting-state FMRI measurements in human brain functional connectomics: A systems neuroscience perspective. *Neuroscience & Biobehavioral Reviews, 45*, 100−118.

Chapter 7

Meditation over the lifespan

The practice of meditation can be found in different age groups and growing evidence has found the promising effects associated with meditation practices across different age groups, from children to old adults. Given that different age groups have different needs and goals for meditation practice, and that people in different developmental stages have distinct physical and mental characteristics such as brain and physiology as well as social behavior, it remains largely unknown how to develop specific meditation programs to meet and satisfy these practitioners during different developmental stages. Moreover, for children, adolescents, adults, and older adults, implementing appropriate meditation programs in education, workplaces, and retirement settings would be critical for meeting their needs. We first discuss early development of attention and self-regulation in children and adolescents mainly in the K-12 school setting. Second, we focus on young and middle-aged adults who are in universities and workplaces and are often more stressed than other age groups. Third, we discuss older adults who are usually in retirement, but have treated health and well-being as priorities in daily life. We also discuss potential meditation programs for different populations based on current evidence.

Development of self-regulation and attention networks

Self-regulation or self-control is defined as the capacity of regulating one's emotions, cognition, thoughts, and behavior in order to achieve positive goals and outcomes. For instance, self-regulation is critical for maintaining good relationships, achieving academic success, and sustaining individual health and well-being. Therefore self-regulation is vital for human survival and adaptability (Tang, 2017). Research has shown that self-regulation is supported by specific brain networks involved in attention (Posner, 2012). In general, attention includes three components——obtaining and maintaining the alert state, orienting to sensory stimuli, and resolving conflict among competing responses (Petersen & Posner, 2012). The alerting network is modulated by the brain's norepinephrine system and involves major nodes in the frontal and parietal cortices. The alert state is critical to high-level performance because changes in alertness lead to a rapid change from a resting

The Neuroscience of Meditation. DOI: https://doi.org/10.1016/B978-0-12-818266-6.00008-3

state to one of increased receptivity to the target. The orienting network interacts with sensory systems to target stimuli and amplify information relevant to task performance. Most importantly, research has shown that the orienting network exerts much of the regulatory control during infancy and early childhood (Posner, Rothbart, Sheese, & Voelker, 2012; Rothbart, Sheese, Rueda, & Posner, 2011). The executive network is involved in resolving conflict and is done both by enhancing activity in networks related to our goals and inhibiting activity in conflicting networks. These types of controls are affected by long connections between the nodes of the executive network and cognitive and emotional areas of the anterior (e.g., frontal) and posterior portions of the brain. One key brain area——the anterior cingulate cortex (ACC)—is the core node of executive attention network and self-regulation. Therefore the executive network is important for voluntary control and self-regulation (Bush, Luu, & Posner, 2000; Sheth et al., 2012).

During early development, attentional orienting operates in conjunction with the actions of caregivers (e.g., parents) to provide regulation of behavior. This association underlies the frequent observation that infants and young children are largely controlled by their environment, including their caregivers. As we get older, the executive attention network comes to dominate self-regulation allowing internal goals to guide our behavior. Using parent report questionnaires (Rothbart, 2011), self-regulation is assessed by effortful control (EC), a higher-order temperament factor. In childhood, performance on conflict-related cognitive tasks is positively related to measures of children's EC (Rothbart, 2011). During childhood and in adulthood, EC is correlated with school performance and indices of life success, including health, income, and successful relationships (Checa & Rueda, 2011; Moffitt et al., 2011). These findings may shed light on the training target of self-regulation through attentional orienting networks in children and executive attention network in adults, respectively.

Brain networks of self-regulation during development

In neuroimaging literature, resting state refers to the state of the brain when there is no task at hand and individuals are simply resting in the scanner. Resting-state scans can be applied at any age, therefore it has become an important tool to study the brain networks in development (Raichle, 2009). One of the brain networks activated during resting state is the executive attention network involved in resolving conflict and is related to parent reports of EC (Dosenbach et al., 2007; Fair et al., 2009). Studies have examined how brain networks change with age and the results suggested that during infancy and early childhood most brain networks involve short connections between adjacent areas, but the long connections important for self-regulation develop slowly over childhood (Fair et al., 2009; Gao et al., 2009). However, one study suggested that findings involving short connections in children may be due to motion-related artifacts that were not adequately removed in these studies. Moreover, after removing high-motion frames, a more adult-like brain network architecture emerges (Power, Barnes, Snyder, Schlaggar, & Petersen, 2012). Further research is needed to fully investigate and validate these results. However, we do often witness a development in self-control as children age, and achievements in self-regulation have often reported between infancy and 7−8 years of age (Rueda et al., 2004). A large functional magnetic resonance imaging (fMRI) study for participants from 4 to 21 years showed a relationship between the ability to resolve conflict in a Flanker task and the size of the right dorsal ACC in the early years of childhood as well as the connectivity of the ACC in later years. These findings are intriguing because characteristics of the ACC are related to impulse, attention, and executive problems in neurodevelopmental disorders, indicating a neural foundation for self-regulation abilities along a continuum from normality to pathology (Fjell et al., 2012). The ACC appears to be a brain target of training self-regulation and we could use methods or techniques such as meditation to strengthen the ACC's activity and connectivity that may, consequently, induce positive behavior changes and outcomes (Tang, 2017; Tang, Holzel, & Posner, 2015).

In longitudinal studies, the ACC of 7-month-old infants was activated when they detected an error (Berger, Tzur, & Posner, 2006), suggesting that they have preliminary executive attention in place, even though parents are not yet able to report on EC and infants have yet to carry out instructed behaviors. It was not until the age of 3 years that children began to show regulation of their behavior by slowing their next response following an error, as adults do (Jones, Rothbart, & Posner, 2003). Another longitudinal study was conducted on the development of self-regulation during infancy and childhood. The testing began when the infants were 7-months old. Because infants are not able to carry out voluntary attention tasks, a series of attractive stimuli were presented on a screen in a repetitive sequence (Clohessy, Posner, & Rothbart, 2001). Infants oriented to them by moving their eyes (and head) to the location. On some trials infants showed they anticipated what was coming by orienting prior to the stimulus. Results showed that infants who made the most anticipatory eye movements also exhibited a pattern of cautious reaching toward novel objects, which predicts EC in older children. In addition, infants with more anticipatory observation skills showed more spontaneous attempts at self-regulation when presented with somewhat frightening objects (Rothbart, 2011; Sheese, Rothbart, Posner, White, & Fraundorf, 2008). When the children were at the age of 18−20 months, they were retested and genotyped, and then retested again around the age of 4 as they were able perform the attention network test (ANT) as a measure of executive attention. The early regulatory effects in infancy and at age 2 were correlated with their later orienting network scores rather than executive network performance in the ANT. In addition, in infancy, orienting of attention was related to lower negative and higher positive affect. But by age 2 years, orienting was no longer related to affect. Later in childhood and for adults, EC is related to lower negative affect (Rothbart, 2011). These findings indicate that the orienting network provides primary regulatory function during infancy. The orienting network continues to serve as a control system, but starting in childhood executive attention appears to dominate in regulating emotions and thoughts (Isaacowitz, 2012; Posner et al., 2012; Rothbart et al., 2011). This parallel use of the two networks are in line with the findings that in adults the frontal-parietal network controls task behavior at short time intervals whereas the cingulo-opercular network exercises strategic control over long intervals (Dosenbach et al., 2007). In summary, cognitive and emotional control systems seem to arise as part of attentional networks.

Meditation in young children

Given that children aged 4 years or older could perform tasks to measure executive attention, studies of training self-regulation using a computerized or meditation program often focus on this age group. There are two types of training methods: (1) network training that targets a certain brain network using repeated cognitive tasks such as attention, working memory, or learning; and (2) state training which aims to change brain (and body) states without using repeated tasks such as meditation and physical exercise (Tang & Posner, 2009, 2014). We take both training methods as examples to demonstrate their similarities and differences in children.

During childhood the ability to resolve conflict is correlated with parental reports of their children's ability to regulate behavior, suggesting the association between executive network and behavioral self-regulation. Previous studies have shown the training effects of attention in children between the age of 4−7 years using computerized programs that involved specific practices engaging executive attention. For example, five days of attention training produced improved networks underlying executive attention and improved intelligence quotient scores. After training, The N2 component of electroencephalogram (EEG) showed clear evidence of improvement in network efficiency in resolving conflict as the result of training (Rueda, Rothbart, McCandliss, Saccamanno, & Posner, 2005). The N2 component is one of averaged electrical potentials which has been found to arise in the ACC and is related to the resolution of conflict. These data suggest that training altered the executive network for the resolution of conflict in the direction of being more like in adults (Posner, Rothbart, & Tang, 2013; Rueda et al., 2005). A number of studies with varying training methods have been shown to improve executive attention and function in preschool children (Diamond, Barnett, Thomas, & Munro, 2007; Diamond & Lee, 2011; Stevens, Lauinger, & Neville, 2009). These practices may be part of classroom activities or individual computer training, and may involve attention or working-memory tasks (Rueda, Checa,

& Combita, 2012; Rueda et al., 2005). All of these methods have involved what we call attention training (one area of network training), that is, the deliberate exercise of particular brain networks related to attention functions (Tang & Posner, 2009). Of note is that programs that teach specific skills to preschool children often do not maintain their advantage and effects when compared to experimental and control groups (Heckman, 2006). However, a few skills continue to show an advantage after training, such as skills involving the ability to employ aspects of self-control (Ludwig & Phillips, 2008; Moffitt et al., 2011). One explanation may be that self-control is influenced by the executive attention network, and training effects of executive attention network seem to be maintained (Posner et al., 2013).

How about state training on executive attention and self-regulation in preschool children? One mindfulness meditation, integrative body—mind training (IBMT), has been developed and tested in multiple randomized controlled trials (RCTs) in populations from 4 to 90 years of age. Our work suggests that even short-term IBMT improves executive attention and self-control, increases positive emotion, reduces stress of cortisol, and improves immune function through central (brain) and autonomic (body) nervous system interaction (Tang, 2009, 2017; Tang et al., 2007). It may be thought that preschool children would not be able to be trained in meditation that requires strict adherence to instructions. To test whether young children could benefit from short-term IBMT, we conducted a pilot intervention study with 60 preschool children (aged 4—5 years) following 10 h of IBMT within 1 month. As depicted in Fig. 7.1, our findings showed significant increases in the children's self-control scores [attentional focusing, inhibitory control (IC), perceptual sensitivity, and EC] measured by the children's behavior questionnaire, and in their executive function on two observed executive function tasks compared to an active control condition. Moreover, these IBMT effects were maintained for up to 15 months later. We also randomly selected 20 children (10/group) to examine event-related potential (ERP) using the go/no-go task and found that IBMT significantly changed the amplitude of the ERP component P3 thought to involve the ACC among children in the IBMT condition, but not in the control condition, suggesting IC improvements among IBMT children. These preliminary results suggested that young children could learn and benefit from meditation and the brain results using EEG and ERP measures suggested potential different mechanisms of meditation from those of computerized cognitive training (Tang, 2009; Tang, Yang, Leve, & Harold, 2012). These findings are consistent with brain structural changes following network training or state training. For example, poor readers aged 8—10 years received 100 h of intensive cognitive training—a remedial instruction program to develop reading skill. Prior to training, poor readers had lower fractional anisotropy (FA) (an index of white-matter integrity) than good readers in the left

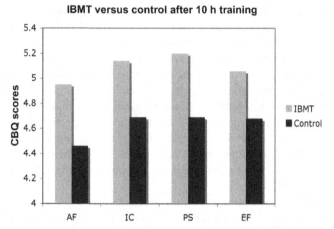

FIGURE 7.1 Children's self-control scores measured by children's behavior questionnaire (CBQ) after 10 h IBMT or control training. *AF,* Attentional focusing; *IC,* inhibitory control; *PS,* perceptual sensitivity; *EC,* executive function.

anterior centrum semiovale. After training, increased FA in white matter was found in the same region. The FA increase was correlated with increased myelination and improvement in phonological decoding ability (Keller & Just, 2009). In contrast, about 5−10 h of state training IBMT increased FA in white matter in several regions such as anterior and superior corona radiata, corpus callosum. White-matter changes were correlated with improved emotion regulation. However, while 5 h of IBMT only increased axonal density, 10 h of IBMT increased both axonal density and myelination (Tang et al., 2010, 2012), suggesting network training and state training work in different mechanisms.

Clearly, 4-year old children know limited words and could not understand and follow exact meditation instructions to practice IBMT, so how could these children learn and grasp meditation? Given the limited evidence, we don't know exactly the underlying mechanisms, but one possibility may be that young children use different learning strategies such as implicit and statistical learning rather than the explicit learning of adults, consistent with the "less is more hypothesis" (Newport, 1990; Saffran, Aslin, & Newport, 1996). Another possibility is that IBMT is a state training program that emphasizes state induction rather than script instructions used in conventional teaching and learning. We also modified IBMT by using children-friendly cartoons or stories to create an environment and learning context that could help children enter the meditative state more naturally (Tang, 2009; Tang et al., 2012). However, further research is warranted to understand and evaluate the effects of meditation in children.

At what age can children practice meditation?

Problems associated with deficiencies in self-control are common and include increased risk of school failure, attention deficit hyperactivity disorder, depression, antisocial behaviors, and drug abuse. Previous findings have indicated that childhood is a critical time for the development of self-control, with long-term and cascading effects for children who show deficits in this area. For example, in a longitudinal study of 1000 children, self-control problems in 3-year-old children were associated with poorer physical health, drug dependence, financial difficulties, and criminal offenses 29 years later (Moffitt et al., 2011). These studies on network training (e.g., cognitive training) and state training (e.g., meditation) have consistently suggested that children aged 4 years could benefit from both types of training. This convergent evidence indicates that this specific age group may be the youngest group for meditation practice or other mental training to be effective. Thus it is of significant public health benefit to improve children's self-control capacities early in development in order to prevent the cascade of risky behaviors and problems. Moreover, some brain networks are linked to more specific sensory and motor aspects of learning, others are more general, playing a role in a wide variety of tasks such as reading and computing. We have discussed three attention networks——alerting, orienting, and executive— that are involved in school and learning (Posner, 2012). Being able to carry out successful learning in school depends on the efficiency of the alerting network. The orienting network exerts most of the regulatory control in infancy and early childhood and plays a role in the classroom by reducing distraction and amplifying input relevant to the subject being studied. However, the age of 4 years seems to be a critical developmental time for switching self-regulation from the orienting network to the executive network. The executive network is important for self-regulation in middle childhood and beyond and its efficiency correlates with school performance (Checa & Rueda, 2011; Posner et al., 2013). The executive network is involved in resolving conflict and also serves as a means of self-regulation through control of brain networks involved in emotion, cognition, and behavior. Meanwhile, at the age of 4, children are also sensitive to brain changes following either cognitive training or meditation (Posner et al., 2013). Taken together, appropriate training programs developed specifically for children aged 4 and older would be imperative for enhancing self-control and later-life outcomes.

Meditation in children and adolescents

Different age groups have different needs from meditation practice. After the age of 4, children progress further into childhood and adolescence (usually associated with the teenage years) and are often in K-12 education and school systems. Many studies have reported the implementation and effects of meditation programs in education settings during this developmental period (Flook, Smalley, & Kitil, 2010; Parker, Kupersmidt, Mathis, Scull, & Sims, 2014; Tang, 2017; Tang, Tang, Jiang, & Posner, 2014; Tang et al., 2012).

There has been nearly a decade of research work and effort in teaching meditation to school children and the results seem promising. In recent years the British government and private funding agencies have supported several large research projects across the country to explore how to integrate evidence-based mindfulness meditation programs into the school system. For example, the Anna Freud National Centre for Children and Families in partnership with University College London are conducting one of the largest mental-health trials in up to 370 schools in the United Kingdom to investigate whether children could benefit from mindfulness exercises, relaxation techniques, and breathing exercises that aim to help them regulate their emotions. Over 5000 trained classroom teachers deliver mindfulness training in UK schools. Although the study will run until 2021, the preliminary results in primary and secondary schools are promising——mindfulness meditation seems to help promote mental health and resilience.

Adolescence is a critical time of change and development. Learning skills to build resilience has the potential to help adolescents navigate these challenges at school and apply these skills throughout their lives. To explore how to prepare adolescents to improve resilience and manage their emotional health, Oxford, Cambridge, and University College London initiated a large randomized control trial——My Resilience in Adolescence (MYRIAD) project, a $10 million Strategic Award from the Welcome Trust. The MYRIAD project (http://myriadproject.org/schools/) focuses on the comparative value of mindfulness-based interventions to support resilience and well-being in youths. Recruitment from 84 UK schools has been completed. Researchers are comparing two programs——good quality social emotional learning taught in schools using a computer program (known as teaching as usual) versus mindfulness techniques (a way of being present to experiences as they happen rather than worrying about what has happened or might happen in the future). Given that adolescence is a transitional stage of physical and psychological development, this population may benefit from meditation across diverse areas such as cognitive, affective, social, brain, physiological, and behavioral domains (Tang, 2017). Studies have shown that meditation-based interventions in the classroom can impact a wide range of indicators of positive psychological, social, and physical well-being and flourishing in children and young people. Regular practice helps improve mental health and well-being, self-regulation, academic learning, positive behavior, and self-esteem (Parker et al., 2014; Tang et al., 2014; Weare, 2013; Zenner, Herrnleben-Kurz, & Walach, 2014). Since half of all mental illnesses begin by the age of 14, it is crucial to provide evidence-based techniques such as meditation to students to improve self-regulation and prevent mental illness in advance (Tang, 2017).

US researchers have also explored the applications of meditation in schools. In second and third grade children aged 7—9 years, an 8-week RCT of school-based mindful awareness practice was evaluated in comparison with a reading program as the control condition. The children's executive function was assessed based on questionnaires teachers and parents completed immediately before and after the 8-week period. Results showed that children in the mindful group with lower executive function received greater improvement in executive function, behavioral regulation, and metacognition compared to the reading control group (Flook et al., 2010). These results may suggest an effect of mindfulness on children with executive-function difficulties. Since teachers and parents reported changes, improvements in children's behavioral regulation may be generalized across settings. Future work is warranted using neurocognitive tasks of executive functions, behavioral observation, and multiple classroom samples to replicate these preliminary results. Another cross-sectional study aimed to assess the feasibility and effectiveness of a mindfulness education in a substance abuse prevention program for fourth and fifth grade children in elementary schools. The goal of the research was to investigate whether children in the intervention group who received 4-week meditation training in the classroom would show enhanced self-regulation and respond effectively to stress and prevent poor decision-making on alcohol and tobacco use. Students completed self-reports of their intentions to use substances and an executive function task at baseline and postintervention. Teachers also rated students on their behavior in the classroom. Findings showed that compared to the wait-list control, the intervention group had significant improvements in executive functioning skills, and a marginally significant increase in self-control abilities (boys only). In addition, significant reductions were found in aggression and social problems, and anxiety (girls only). No significant differences across groups were found for intentions to use alcohol or tobacco. Despite the relatively small sample size, these results suggest that mindfulness education may be beneficial for increasing self-regulatory abilities, which is important for substance-abuse prevention (Parker et al., 2014). Relatedly, an ongoing RCT study in Finland is exploring the effectiveness of a 9-week, school-based, mindfulness program compared to a relaxation program in a nonclinical children and adolescent sample. The study has three components——mindfulness group, active control relaxation group, and waitlist group—with a sample of about 3000 students aged 12—15 years. This study focuses on the effects of mindfulness meditation on different aspects of mental well-being such as resilience, existence/absence of depressive symptoms, experienced psychological strengths and difficulties, cognitive functions, psychophysiological responses, academic achievements, and motivational determinants of practice. Students, teachers, and parents complete questionnaires of mental well-being before and after the intervention, and they will all be followed for 6—12 months (Volanen et al., 2016). Although the study is still ongoing, based on previous research, it seems promising that mindfulness meditation will outperform (RT) and the waitlist control group (Tang et al., 2007, 2015).

Improved self-regulation and attention appear to have a beneficial effect on learning school subjects such as literacy and numeracy (Checa & Rueda, 2011; Posner & Rothbart, 2014). In one study, higher self-regulation in self-reported temperament scales was associated with better school performance, while better performance on the executive attention network was related to improved mathematics performance in particular (Checa & Rueda, 2011). Our RCT study using five sessions of mindfulness meditation (the IBMT) found improved executive attention which, in turn, is correlated with self-regulation. However, our prior work did not examine whether a brief IBMT intervention would improve the academic performance of adolescents in schools. To address this research gap, we randomly assigned 208 middle and high school students into IBMT or RT groups (104 in each group). They received 6 weeks of IBMT or RT intervention (20−30 min per session from Monday to Friday; a total of 10 h) during school lunch break prior to their yearly final examinations. Given that self-regulation involves important components of attention control and emotion regulation, we thus measured the efficiency of three attentional components using the ANT. We also assessed the mood state using the profile of mood states (POMS), fluid intelligence using the Raven's standard progressive matrix, and self-reported stress using the perceived stress scale (PSS). All of these are frequently used measures of performance or subjective experience (Tang et al., 2007). The school provided information of the official grades obtained at the end of the academic year by each student. Academic performance grades were obtained for literacy (Chinese), mathematics, and second language (English). Before training, two groups of IBMT and RT did not show any significant differences in assays and grade scores. After training, we examined differences in ANT, Raven's Matrices, POMS and PSS, and grades between IBMT and RT groups. We found that 10 h of IBMT significantly improved executive and alerting attention, indicating greater self-control and sustained attention ability. However, orienting attention was only marginally significant following IBMT. The same amount of RT also improved all attention efficiency, but not significantly. We also tested whether intelligence improved after training and detected a significant improvement in Raven scores following short-term IBMT (but not RT). Because the efficiency of executive attention improved, we expected better self-regulation of emotion. After training, there were significant differences in the IBMT group (but not the RT group) in the POMS scales: anger−hostility (A), depression−dejection (D), F (fatigue−inertia), tension−anxiety (T), and vigor−activity (V). These results indicated that short-term IBMT can enhance positive moods and reduce negative ones. Meanwhile, PSS also favored the IBMT group after training. The academic achievement indexed by the final mean of grades of literacy, mathematics, and second language (English) was also significantly improved following

IBMT. After training, we examined the behavioral and grade changes between the two groups and found significances in the IBMT group compared to the RT group. Taken together, short-term meditation can improves self-regulation and academic performance in adolescents (Tang et al., 2014).

A meta-analysis of first to 12th grade students summarized the evidence regarding the effects of school-based mindfulness interventions on psychological outcomes and suggested that overall effect sizes were about Hedge's $g = 0.40$ between groups and within groups. Between group effect sizes for domains included cognitive performance $g = 0.80$, stress $g = 0.39$, and resilience $g = 0.36$, which were all significant, while emotional problems and third person ratings were not (Zenner et al., 2014). These results may suggest mindfulness meditation in children and youths as promising, particularly in improving cognitive performance and resilience to stress. However, the results of emotional problems were inconsistent, probably reflecting the diversity of study samples, variety in implementation and practices, and the wide range of instruments used. A careful examination and replication is warranted. A recent review utilized the 2012 National Health Interview Survey data to investigate how complementary and alternative medicine (CAM) such as meditation helps mental health issues in children, a serious public health concern in the United States. Based on the large sample ($N > 10,000$), evidence suggests that CAM use is more popular among children with mental health issues compared to those without ($\sim 20\%$ vs 10%). The main reasons for children to use CAM are because they are helpful ($\sim 70\%$), natural ($\sim 55\%$), and holistic ($\sim 45\%$). Predictors of CAM use are children who are female, whose parents had a higher educational level and socioeconomic status, and who had at least one comorbid medical condition. Unfortunately, only $\sim 18\%$ of CAM usage was recommended by medical doctors. Given the low rate of CAM use recommended by doctors, these results may suggest medical professionals need to get more exposure to CAM knowledge and experience, which could help improve patient–physician communication and understanding with regard to the use of CAM as complementary treatment approaches. Moreover, as families with lower educational level and socioeconomic status seldom use CAM, strategies designed to increase the awareness of CAM and reduce disparities in accessing CAM are warranted. Overall, meditation has become a popular and acceptable training or intervention for children and adolescents (Wang, Preisser, Chung, & Li, 2018).

Meditation in young and middle-aged adults

In contrast to children and adolescents, young and middle-aged adults have different needs and challenges in schools and workplaces. According to the annual nationwide survey conducted by the American Psychological Association to examine the state of stress across the country, young and middle-aged adults (18–38 years old) are more stressed than other age groups, and economy was most likely to be cited as a cause of stress in this age group (~40%). Therefore stress reduction and management in work-places seem critical for this group. One unhealthy coping mechanism—smoking—is reported by 14% of adults. However, others are taking healthier actions to deal with stress (e.g., 53% of Americans exercise or take part in physical activity to cope with stress). One coping method is on the rise, with 12% of people using meditation or yoga to manage their stress. Many Americans understand that emotional support is crucial to dealing with the stress in their lives. Nearly 75% feel they have to find someone they can rely on for emotional support. However, if people could not find an appropriate individual, could they handle stress by themselves? Meditation seems to be a very promising method based on growing scientific evidence (American Psychological Association, 2017).

In our series of RCTs, we have reported that compared to RT, even brief IBMT can improve self-regulation and executive attention, immune function, as well as positive moods. Brief IBMT also reduces negative moods such as anxiety, depression, anger, and fatigue and decreases stress-related cortisol concentration (Tang, 2017; Tang et al., 2007, 2015). For example, in young adults, five 20-min sessions of IBMT significantly reduces stress in subjec-tive self-reports of stress and objective cortisol concentration of stress hor-mone and immunoreactivity (Tang et al., 2007). These findings suggest that IBMT could be an effective way to help the most stressful age group (18–38 years old) to better handle stress.

If one practices IBMT over five sessions, does continued training show dose dependence? In two RCTs, we randomly assigned young adults into IBMT or RT for 20 sessions within 4 weeks (~10 h in total). With this longer period of training, greater changes in aspects of attention were found——both alerting and executive networks were more improved with IBMT than with RT. The increased dose of IBMT significantly reduced cor-tisol concentration and significantly increased immunoreactivity even at baseline before any cognitive challenge (one stressor). These suggest that subjects with only 4 weeks of meditation training were able to cope with the stress of everyday life to a greater degree than they were able to before train-ing. Moreover, subjects had better immune function following 4 weeks of meditation training (Fan, Tang, Ma, & Posner, 2010; Fan, Tang, & Posner, 2014). Lower stress and higher immune-function levels at baseline indicate the benefits of meditation in improving health and well-being (Tang, Tang, & Gross, 2019).

In the workplace, burnout and work-related stress have become a major concern in our society. Job burnout is a state of physical or emotional exhaustion that also involves a sense of reduced accomplishment and loss of personal identity, which can happen in any profession. For example, recent evidence has suggested that even in healthcare professionals (HCPs), burnout and stress are growing concerns, as they can induce a wide range of physical and psychological symptoms that can affect the quality of patient care. Mindfulness meditation may be a good approach to address burnout in HCPs. A study used within-subjects design to investigate the effects of 8-week mindfulness course before, after, and during follow-up in HCPs. Results suggested significant reductions in two subscales of burnout—depersonalization and emotional exhaustion. Three facets of dispositional mindfulness also showed significant increases at follow-up. Although the sample size is small ($N = 18$), mindfulness meditation is helpful for HCPs. Future research is warranted with greater sample size and more rigorous research design (Ellen Braun, Kinser, Carrico, & Dow, 2019). One recent review examined whether mindfulness meditation improves the well-being of HCPs in burnout, distress, anxiety, depression, and stress and concluded that, in general, meditation was associated with positive outcomes in most of the measures except burnout (Lomas, Medina, Ivtzan, Rupprecht, & Eiroa-Orosa, 2019). Another recent meta-analysis aimed to address the limitation of previous meta-analyses, including the narrow scope regarding intervention type, target population, and measures. This meta-analysis (Spinelli, Wisener, & Khoury, 2019) used a more comprehensive and rigorous examination for HCPs and suggested that mindfulness meditation had a significant moderate effect on anxiety, depression, psychological distress, and stress (Hedge's $g = 0.41 - 0.52$). Additionally, for burnout and well-being at postintervention, small to moderate effects were found but effects were not significant for physical health and performance. In summary, mindfulness meditation is fairly effective in reducing distress and improving well-being in HCPs and other professionals. However, meditation techniques also matter and it is important to know participants' needs prior to selecting the type of meditation. Future studies should assess differences in meditation techniques and include active control conditions to further evaluate the effects of meditation on psychological health.

Prolonged stress without effective coping often leads to mental disorders such as depression and anxiety. We will discuss depression as an example since individuals with depression often show deficits in self-control of attention and emotion. Mindfulness meditation has been shown to improve self-control ability through strengthening brain activity and connectivity of the ACC, prefrontal cortex (PFC), and striatum (Tang et al., 2009, 2010). We conducted a pilot study to determine the efficacy of short-term meditation using IBMT in adults with first episode depression and its brain mechanism. Thirty-three first episode depression and 33 matched healthy adults without any treatment history participated in this study. Subjects received 10 h of IBMT in total, in twenty 30 min/session over 4 weeks. Before and after intervention, subjects completed the attention network test and POMS to assess changes in self-control. Brain measurement of cerebral blood flow (CBF) was used to detect brain changes using GE 3-head SPECT scanner. Before intervention, compared to healthy controls, the depression group showed significant deficits in executive attention (see Fig. 7.2, depression vs healthy controls, $**p < .01$) and greater anger, depression, fatigue, anxiety, and confusion. Furthermore, CBF showed global reductions especially in ACC and adjacent PFC, insula and striatum, associated with self-control and reward. After intervention, alerting and executive attention (see Fig. 7.3, depression group post-IBMT vs pre-IBMT, $**p < .01$), mood indexes, and the CBF of the above regions showed significant increases. These findings suggest that brief mindfulness meditation has the potential to alleviate symptoms of mood disorders through improved regulation of self-control and reward networks in the brain (Tang, 2017).

Relatedly, we also conducted an RCT to examine whether 2 weeks of IBMT (30 min per session, 5 h in total) could reduce smoking and craving. Our results were in line with our hypothesis that IBMT improves self-control network in the ACC and adjacent PFC. Further, it also changes addiction

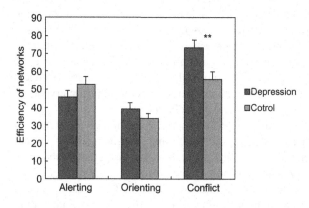

FIGURE 7.2 ANT in depression vs healthy controls before intervention. Executive attention showed significance, $**p < .01$.

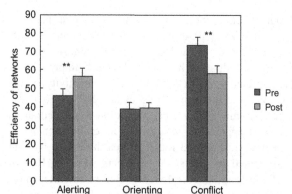

FIGURE 7.3 ANT in depression group after 10 h of IBMT intervention. Alerting and executive attention showed significance, $**p < .01$.

behavior and moods, and reduces stress (Tang, Tang, & Posner, 2013). It should be noted that not only meditation changes brain function, but also brain structure in white-matter connectivity and gray matter through myelination, axonal density, and volumetric increases (Tang et al., 2015). These brain plasticity measures also accompany physiological changes which could support behavioral changes and treatment of disorders (see Chapter 5: Sympathetic and parasympathetic systems in meditation, and Chapter 6: Brain regions and networks in meditation). Taken together, meditation practice strengthens self-control and helps prevent and treat mental disorders related to deficits in self-control in adults (Tang, 2017).

Meditation in older adults

Older adults have different mindsets, attitudes, needs, and lifestyles than young and middle-aged adults. Compared to young and middle-aged adults with the highest reported stress levels on average, older adults in three age groups (39–52, 53–71, and 72 + years) have the lowest stress levels among all age groups. However, health care is most likely to be a source of stress for this group, which motivates them to pay more attention and effort into maintaining health and well-being (American Psychological Association, 2017). Many studies have reported positive effects of meditation in the healthy aging population and patients with neurodegenerative disorders such as dementia and Alzheimer's disease. A review summarized the impact of mindfulness meditation on three areas of functioning in older adults: behavioral and neural correlates of attention, psychological well-being, and inflammation (Fountain-Zaragoza & Prakash, 2017). It has been suggested that improved focused attention may mitigate the decline of attention-control abilities in this group and allow older adults to utilize their preserved emotion regulation abilities. Evidence shows some improvements in attention control in older adults, but some studies suggest no benefits in performance. Additionally, it is suggested that meditation enhances psychological and physical aspects of well-being, accompanied by improved inflammation. Therefore it is important for future research to use RCTs and active control conditions to establish the causality of practice effects on these dimensions in older adults (Fountain-Zaragoza & Prakash, 2017).

We conducted an RCT to compare two popular interventions——physical exercise and mindfulness meditation in aging population. Previous studies have shown that physical exercise and mindfulness meditation lead to increased physical and mental health. However, it is unknown whether the two practices share the same underlying mechanisms. In the same local communities, we randomly chose healthy and high-functioning Chinese older adults from a longitudinal national health project based on their willingness to participate in IBMT or physical exercise groups. Each group had practiced daily in the local community for an average of 1 h. The IBMT or exercise instructor in the same community supervised the practice sessions. All participants were living independently and were matched in age, sex, education, and health status. At the end of the 10 years of training, we measured brain activity, physiology, and behavior. No sedentary control group was assigned because such a group would be unlikely to maintain good health over a 10-year period so that our data represent only differences between mental and physical training. In comparison with exercise, we found significantly higher results for IBMT on (1) ratings of quality of life, (2) parasympathetic activity indexed by high-frequency heart rate variability, (3) basal Secretory Immunoglobulin A (sIgA) levels, and (4) basal cortisol levels. The IBMT group also showed stronger brain connectivity between the dorsal ACC and the striatum using resting-state fMRI and greater gray matter in the striatum using structural MRI than the exercise group, suggesting increased brain plasticity in aging population. The exercise group showed lower overall heart rate and greater chest respiratory amplitude than the IBMT group at baseline (see Chapter 5: Sympathetic and parasympathetic systems in meditation). Our results suggested that meditation and physical exercise may work in part by different mechanisms, with exercise producing greater physical fitness and meditation producing greater central nervous system changes. These findings suggest that combining physical and mental training may lead to better health and quality of life during aging (Tang, 2009; Tang, Posner, & Rothbart, 2013).

Finally, among older adults, pain, neurodegenerative, and psychiatric disorders are common. Meditation has been widely utilized to alleviate physical pain, as well as for older adults who have mild cognitive impairment, cognitive decline, dementia, and AD. There is some evidence suggesting that regular meditation, physical exercise, or other integrative health programs could help improve daily functioning, emotion and stress management, and health maintenance (Ahn & Hyun, 2019; Chouinard, Larouche, Audet, Hudon, & Goulet, 2019; Gard, Hölzel, & Lazar, 2014; Malinowski, Moore, Mead, & Gruber, 2017; Russell-Williams et al., 2018). However, these benefits of meditation in older adults require large-sample RCTs with active controls to fully validate and replicate the results.

Is there a critical period for meditation learning and practice?

Meditation is the systematic training of attention and self-control, and childhood is a critical period for the development of self-control, with long-term and cascading effects for children who show deficits in this area. Training self-control during childhood through meditation seems to be a good choice and may prevent future potential risks such as learning difficulties, school failure, problems in the workplace and relationships, as well potential health issues and disorders. Although some preliminary results using child-friendly versions of meditation programs including our work have been promising, more scientific evidences are warranted.

Adolescents and young adults are sensitive to environmental influences as their brains are still developing. Previous studies have suggested that problems associated with self-control deficits are common in this group and often lead to increased risk of antisocial behavior and mental disorders such as attention deficit disorder, depression, anxiety, and drug abuse. For example, many studies indicate the peak prevalence of substance use disorders (SUDs) occurs during school years and students show heightened polysubstance use (e.g., alcohol use, smoking, cannabis) compared to the nonattending peers. Other populations (e.g., middle-aged or older adults with SUDs) most likely began engaging in problematic substance use during late adolescence/early adulthood (i.e., college years), suggesting that this is a critical period for intervention. Therefore the heightened risk of polysubstance use in students has become a widespread critical public health issue. Effective intervention such as meditation for this population could be important and even more impactful (in terms of reducing potential years of life lost and chronic impairment) compared to targeting older populations. Most importantly, meditation practice can strengthen brain networks of self-control and improve self-control abilities to better handle stress, academic performance, and behavioral problems in schools. Therefore it can be highly beneficial for this group to learn meditation and then integrate it into their lifestyle. Indeed, many basic and translational meditation research studies have focused on this group and growing evidence suggests the feasibility and effectiveness of meditation programs in this population (Tang, 2017). Middle-aged and older adults also benefit from meditation, but adolescence seems to be a critical period for learning and developing this skill, not only for education purposes, but also for character development (Tang, 2017).

It has been suggested that meditation includes at least three components that interact closely to constitute a process of enhanced self-regulation: enhanced attention control, improved emotion regulation, and altered self-awareness (Tang et al., 2015). From the developmental perspective, there is evidence to suggest that early self-regulation is related to more favorable life outcomes. Therefore training and strengthening attention and self-regulation as early as possible could equip children and adolescents better for school, professional and personal performance, and outcomes. Work on meditation with young people has a developing presence and can be effective in promoting a wide range of outcomes. Regular practice has been shown to be capable of improving mental health and well-being, mood, self-esteem, self-regulation, positive behavior, and academic learning. There are many possible promising locations for mindfulness within mainstream education and the health services, including work to improve mental health and well-being for teachers and students, social and emotional learning, and special education (Weare, 2013).

This new knowledge can be used to improve the lives of children and their future. To do this, we need to investigate how to combine methods of improving self-regulation and integrate them into the educational system. Methods involving training specific attention networks (e.g., cognitive training) and those that involve changing brain states (e.g., meditation) may involve different rates of learning and may work through different mechanisms. Thus imaging the brain may give us clues as to how to best develop a combination of training methods. Furthermore, studies on self-regulation may also lead to more refined intervention approaches for improving this critical function. This line of research could provide unusual opportunities for educators, psychologists, and neuroscientists to work together toward the common goal of improving children's lives and their future (Posner et al., 2013; Tang, 2017). Moreover, individual differences in temperament and personality could motivate certain individuals into seeking out meditation and maintaining such practice over the course of their lifespan. As we discussed in Chapter 2, Personality and meditation, personality traits such as openness to experience and conscientiousness are relevant for meditation practices. Interestingly, these two traits are also highly important for psychological health and well-being across the lifespan, suggesting that individuals with high levels of conscientiousness and openness to experience may be the ones that are likely to seek out and maintain practice, thereby leading to better health. These differences should be evaluated and considered in future studies of meditation across all age groups.

In conclusion, as meditation is increasingly becoming a lifestyle practice, its role and effects across development and over lifespan are worth investigating longitudinally, not just limited to a specific age group only for a period of time. In this chapter, we discussed the potential of meditation as mental training for promoting psychological health and well-being, and whether or not it would be suitable for all age groups. Of note, meditation may help us change our mindset and lifestyle——switching from a "doing" culture to a "being" culture, which has particular significance in the digital age.

References

Ahn, H. I., & Hyun, M. K. (2019). Effectiveness of integrative medicine program for dementia prevention on cognitive function and depression of elderly in a public health center. *Integrative Medicine Research, 8*(2), 133–137.

American Psychological Association. (2017). *Stress in America: The state of our nation.* Stress in America Survey. Washington, DC: American Psychological Association.

Berger, A., Tzur, G., & Posner, M. I. (2006). Infant brains detect arithmetic errors. *Proceedings of the National Academy of Sciences of the United States of America, 103*(33), 12649–12653.

Bush, G., Luu, P., & Posner, M. I. (2000). Cognitive and emotional influences in the anterior cingulate cortex. *Trends in Cognitive Science, 4*(6), 215–222.

Checa, P., & Rueda, M. R. (2011). Behavior and brain measures of executive attention and school competence in late childhood. *Developmental Neuropsychology*, *36*(8), 1018−1032.

Chouinard, A. M., Larouche, E., Audet, M. C., Hudon, C., & Goulet, S. (2019). Mindfulness and psychoeducation to manage stress in amnestic mild cognitive impairment: A pilot study. *Aging & Mental Health*, *23*(9), 1246−1254.

Clohessy, A. B., Posner, M. I., & Rothbart, M. K. (2001). Development of the functional visual field. *Acta Psychologica*, *106*, 51−68.

Diamond, A., Barnett, S., Thomas, J., & Munro, S. (2007). Preschool improves cognitive control. *Science*, *30*, 1387−1388.

Diamond, A., & Lee, K. (2011). Interventions shown to aid executive function development in children 4−12 years old. *Science*, *334*(6054), 311.

Dosenbach, N. U. F., Fair, D. A., Miezin, F. M., Cohen, A. L., Wenger, K. K. R., Dosenbach, A. T., ... Petersen, S. E. (2007). Distinct brain networks for adaptive and stable task control in humans. *Proceedings of the National Academy of Sciences of the United States of America*, *104*, 1073−1978.

Ellen Braun, S., Kinser, P., Carrico, C. K., & Dow, A. (2019). Being mindful: A long-term investigation of an interdisciplinary course in mindfulness. *Global Advances in Health and Medicine*, *8*. Available from https://doi.org/10.1177/2164956118820064.

Fair, D. A., Cohen, A. L., Power, J. D., Dosenbach, N. U. F., Church, J. A., Miezin, F. M., ... Petersen, S. E. (2009). Functional brain networks develop from a "local to distributed" organization. *PLoS Computational Biology*, *5*, e1000381.

Fan, Y., Tang, Y. Y., Ma, Y., & Posner, M. I. (2010). Mucosal immunity modulated by integrative meditation in a dose dependent fashion. *Journal of Alternative and Complementary Medicine*, *16*, 151−155.

Fan, Y., Tang, Y. Y., & Posner, M. I. (2014). Cortisol level modulated by integrative meditation in a dose-dependent fashion. *Stress Health*, *30*(1), 65−70.

Fjell, A. M., et al. (2012). Multimodal imaging of the self-regulating brain. *Proceedings of the National Academy of Sciences of the United States of America*, *109*(48), 19620−19625.

Flook, L., Smalley, S. L., Kitil, M. J., et al. (2010). Effects of mindful awareness practices on executive functions in elementary school children. *Journal of Applied School Psychology*, *26*(1), 70−95.

Fountain-Zaragoza, S., & Prakash, R. S. (2017). Mindfulness training for healthy aging: Impact on attention, well-being, and inflammation. *Frontiers in Aging Neuroscience*, *9*, 11.

Gard, T., Hölzel, B. K., & Lazar, S. W. (2014). The potential effects of meditation on age-related cognitive decline: A systematic review. *Annals of the New York Academy of Sciences*, *1307*, 89−103.

Gao, W., Zhu, H., Giovanello, K. S., Smith, J. K., Shen, D., Gilmore, J. H., & Lin, W. (2009). Evidence on the emergence of the brain's default network from 2-weekold to 2-year-old healthy pediatric subjects. *Proceedings of the National Academy of Sciences of the United States of America*, *106*, 6790−6795.

Heckman, J. J. (2006). Skill formation and the economics of investing in disadvantaged children. *Science*, *312*, 1900−1902.

Isaacowitz, D. M. (2012). Mood regulation in real time: Age differences in the roleof looking. *Current Direction in Psychological Science*, *21*(4), 237−242.

Jones, L. B., Rothbart, M. K., & Posner, M. I. (2003). Development of executive attention in preschool children. *Developmental Science*, *6*, 498−504.

Keller, T. A., & Just, M. A. (2009). Altering cortical connectivity: Remediation-induced changes in the white matter of poor readers. *Neuron*, *64*, 624–631.

Lomas, T., Medina, J. C., Ivtzan, I., Rupprecht, S., & Eiroa-Orosa, F. J. (2019). A systematic review of the impact of mindfulness on the well-being of healthcare professionals. *Journal of Clinical Psychology*, *74*(3), 319–355.

Ludwig, J. C., & Phillips, D. A. (2008). The long-term effects of head start on low-income children. *Annals of the New York Academy of Sciences*, *40*, 1–12.

Malinowski, P., Moore, A. W., Mead, B. R., & Gruber, T. (2017). Mindful aging: The effects of regular brief mindfulness practice on electrophysiological markers of cognitive and affective processing in older adults. *Mindfulness (New York)*, *8*(1), 78–94.

Moffitt, T. E., Arseneault, L., Belsky, D., Dickson, N., Hancox, R. J., Harrington, H. L., ... Caspi, A. (2011). A gradient of childhood self-control predicts health, wealth and public safety. *Proceedings of the National Academy of Sciences of the United States of America*, *108*, 72693–72698.

Newport, E. L. (1990). Maturational constraints on language learning. *Cognitive Science*, *14*(1), 11–28.

Parker, A. E., Kupersmidt, J. B., Mathis, E. T., Scull, T. M., & Sims, C. (2014). The impact of mindfulness education on elementary school students: Evaluation of the master mind program. *Advances in School Mental Health Promotion*, *7*(3), 184–204.

Petersen, S. E., & Posner, M. I. (2012). The attention system of the human brain: 20 years after. *Annual Review of Neuroscience*, *35*, 71–89.

Posner, M. I. (2012). *Attention in a social world*. New York: Oxford University Press.

Posner, M. I., & Rothbart, M. K. (2014). Attention to learning of school subjects. *Trends Neuroscience and Education*, *3*, 14–17.

Posner, M. I., Rothbart, M. K., Sheese, B. E., & Voelker, P. (2012). Control networks and neuromodulators of early Development. *Developmental Psychology*, *48*(3), 827–835.

Posner, M. I., Rothbart, M. K., & Tang, Y. Y. (2013). Developing self-regulation in early childhood. *Trends in Neuroscience and Education*, *2*(3-4), 107–110.

Power, J. D., Barnes, K. A., Snyder, A. Z., Schlaggar, B. L., & Petersen, S. E. (2012). Spurious but systematic correlations in functional connectivity MRI networks arise from subject motion. *NeuroImage*, *59*(3), 2142–2154.

Raichle, M. E. (2009). A paradigm shift in functional brain imaging. *Journal of Neuroscience*, *29*(41), 12729–12734.

Rothbart, M. K. (2011). *Becoming who we are: Temperament, personality and development*. Guilford Press.

Rothbart, M. K., Sheese, B. E., Rueda, M. R., & Posner, M. I. (2011). Developing mechanisms of self-regulation in early life. *Emotion Review*, *3*(2), 207–213.

Rueda, M. R., Checa, P., & Combita, L. M. (2012). Enhanced efficiency of the executive attention network after training in preschool children: Immediate and after two month effects. *Developmental Cognitive Neuroscience*, *2*(2), 291.

Rueda, M. R., Fan, J., Halparin, J., Gruber, D., Lercari, L. P., McCandliss, B. D., & Posner, M. I. (2004). Development of attention during childhood. *Neuropsychologia*, *42*, 1029–1040.

Rueda, M. R., Rothbart, M. K., McCandliss, B. D., Saccamanno, L., & Posner, M. I. (2005). Training, maturation and genetic influences on the development of executive attention. *Proceedings of the National Academy of Sciences of the United States of America*, *102*, 14931–14936.

Russell-Williams, J., Jaroudi, W., Perich, T., Hoscheidt, S., El Haj, M., & Moustafa, A. A. (2018). Mindfulness and meditation: Treating cognitive impairment and reducing stress in dementia. *Reviews in the Neurosciences*, *29*(7), 791–804.

Saffran, J. R., Aslin, R. N., & Newport, E. L. (1996). Statistical learning by 8-month-old infants. *Science, 274*(5294), 1926–1928.

Sheese, B. E., Rothbart, M. K., Posner, M. I., White, L. K., & Fraundorf, S. H. (2008). Executive attention and self-regulation in infancy. *Infant Behavior and Development, 31*, 501–510.

Sheth, S. A., Mian, M. K., Patel, S. R., Asaad, W. F., Williams, Z. M., Dougherty, D. D., ... Eskander, E. N. (2012). Human dorsal anterior cingulate cortex neurons mediate ongoing behavioural adaptation. *Nature, 488*, 218–221.

Spinelli, C., Wisener, M., & Khoury, B. (2019). Mindfulness training for healthcare professionals and trainees: A meta-analysis of randomized controlled trials. *Journal of Psychosomatic Research, 120*, 29–38.

Stevens, C., Lauinger, B., & Neville, H. J. (2009). Differences in neural mechanisms of selective attention in children from different socioeconomic backgrounds: An event-related brain potential study. *Developmental Science, 12*, 643–646.

Tang, Y. Y. (2009). *Exploring the brain, optimizing the life.* Beijing: Science Press.

Tang, Y. Y. (2017). *The neuroscience of mindfulness meditation: How the body and mind work together to change our behavior? Springer Nature.*

Tang, Y. Y., Holzel, B. K., & Posner, M. I. (2015). The neuroscience of mindfulness meditation. *Nature Reviews Neuroscience, 16*(4), 213–225.

Tang, Y. Y., Lu, Q., Geng, X., Stein, E. A., Yang, Y., & Posner, M. I. (2010). Short-term meditation induces white matter changes in the anterior cingulate. *Proceedings of the National Academy of Sciences of the United States of America, 107*(35), 15649–15652.

Tang, Y. Y., Ma, Y., Wang, J., Fan, Y., Feng, S., Lu, Q., & Posner, M. I. (2007). Short term meditation training improves attention and self regulation. *Proceedings of the National Academy of Sciences of the United States of America., 104*(43), 17152–17156.

Tang, Y. Y., & Posner, M. I. (2009). Attention training and attention state training. *Trends in Cognitive Sciences, 13*(5), 222–227.

Tang, Y. Y., & Posner, M. I. (2014). Training brain networks and states. *Trends in Cognitive Science, 18*(7), 345–350.

Tang, Y. Y., Posner, M. I., & Rothbart, M. K. (2013). Meditation improves self-regulation over the lifespan. *Annals of the New York Academy of Sciences, 1307*, 104–111.

Tang, Y. Y., Tang, R., & Gross, J. J. (2019). Promoting psychological well-being through anevidence-based mindfulness training program. *Frontiers in Human. Neuroscience, 13*, 237.

Tang, Y. Y., Tang, R., Jiang, C., & Posner, M. I. (2014). Short-term meditation intervention improves self-regulation and academic performance. *Journal of Child and Adolescent Behavior, 2*, 154.

Tang, Y. Y., Yang, L., Leve, L. D., & Harold, G. T. (2012). Improving executive function and its neurobiological mechanisms through a mindfulness-based intervention: Advances within the field of developmental neuroscience. *Child Development Perspectives, 6*(4), 361–366.

Tang, Y. Y., Tang, R., & Posner, M. I. (2013). Brief meditation training induces smoking reduction. *Proceedings of the National Academy of Sciences of the United States of America, 110*(34), 13971–13975.

Volanen, S. M., Lassander, M., Hankonen, N., Santalahti, P., Hintsanen, M., Simonsen, N., & Suominen, S. (2016). Healthy learning mind—a school-based mindfulness and relaxation program: A study protocol for a cluster randomized controlled trial. *BMC Psychology, 4*(1), 35.

Wang, C., Preisser, J., Chung, Y., & Li, K. (2018). Complementary and alternative medicine use among children with mental health issues: Results from the National Health Interview Survey. *BMC Complementary and Alternative Medicine, 18*(1), 241.

Weare, K. (2013). Developing mindfulness with children and young people: A review of the evidence and policy context. *Journal of Children's Services, 8*(2), 141–153.

Zenner, C., Herrnleben-Kurz, S., & Walach, H. (2014). Mindfulness-based interventions in schools—a systematic review and meta-analysis. *Frontiers in Psychology, 5*, 603.

Chapter 8

How to measure outcomes and individual differences in meditation

Scientific investigations stress to be as objective as possible through carefully and thoughtfully manipulating experimental conditions and measuring phenomena via reliable and valid assessment tools. As researchers, we have gone through a long journey of learning and development, both in methods of measurement and analyses to be as objective and accurate as possible in our scientific findings. While increasing conceptual and methodological issues have been uncovered and realized over the past decade, especially due to the problems of reproducibility in research findings, it is important to take note that these issues are common to all research fields and disciplines as people progress in their methods and ways of thinking, and that the research of meditation is no exception. However, the conceptual challenges in the research of meditation are not typical for other research disciplines. For example, psychologists who study memory have a consensus on what memory is and have a clear definition for this construct. For meditation, the definition is not always clear-cut, since the term can refer to a wide range of traditions and techniques, which sometimes makes it extremely challenging for researchers to compare and contrast findings across different studies of meditation (Davidson & Kaszniak, 2015; Tang, Hölzel, & Posner, 2015). Therefore a useful starting point for our discussion on how to measure meditation outcomes would be to go back to what it is that we are studying and investigating when we talk about meditation.

For scientific investigations we first need to succinctly define and describe our investigative target and research question. For meditation research, we are typically interested in the effects of meditation experiences on certain outcomes related to health and well-being. However, this is where the first conceptual issue arises, what meditation practice or technique we are talking about? Meditation, based on what we have defined for the purpose of this book, includes any meditation practices and techniques from ancient traditions such as Hinduism and Buddhism, as well as from contemporary secular training and intervention programs. Although meditation falls under the

The Neuroscience of Meditation. DOI: https://doi.org/10.1016/B978-0-12-818266-6.00009-5

umbrella of contemplative practices, the variety of meditation techniques and traditions available to people and researchers can be overwhelming. Not only are there many potential investigative targets, there are also many ongoing debates about the specific categorization of each technique and tradition. For example, the term mindfulness can often be found in discussions on meditation or vice versa, yet there is no definitive consensus on the definition of mindfulness, or what kind of meditation techniques should be considered as mindfulness practices (Tang, 2017; Van Dam et al., 2018). This conceptual issue of defining what it is that we are studying could be problematic for conducting experiments and interpreting results if it is not adequately addressed. The straightforward solution to this conceptual ambiguity with respect to terms such as mindfulness and meditation is to clearly describe the practice, technique, or tradition that is being studied. More importantly, researchers should refrain from using broad generalization terms such as mindfulness and meditation without a detailed description of the technique or practice (Tang, Jiang & Tang, 2017). For example, the duration and frequency of meditation practice, the delivery format of meditation practice (i.e., face-to-face, online), the amount of total practice time (e.g., examining experienced meditators), and the contexts of meditation practice (e.g., school, workplace, and intervention group) should be described clearly to provide a clear understanding of the meditation practice that is being examined. Although having conceptual agreement on specific categorizations of different meditation traditions and techniques would be useful, the process of coming to an agreement on scientific theories and definitions could take much time and effort. With regard to mindfulness and meditation, the best solution is to avoid drawing broad conclusions and generalizations, but focus on explaining the design, intervention, and results as clearly and accurately as possible to the scientific community and the general public. Readers could then make a judgment themselves about the meditation practice without being confused or misguided into thinking that it is something else.

In addition to conceptual ambiguity, another important knowledge gap that some meditation researchers are missing is the actual practice experience of meditation, which can not only impede the progress of coming into consensus on complicated theoretical terms relate to meditation, but also can increase the likelihood of misinterpreting or misunderstanding scientific results. Unlike other mental processes such as memory or emotions, meditation is not something that we engage in, or experience, every day. Thus without ever having a real meditation experience or achieving a meditative state, it is extremely difficult to infer the experience purely based on written description or scholarly definition. This lack of actual experience could posit hurdles when delineating the conceptual and practical similarities and differences among a wide range of meditation traditions and techniques as there are fine-grained nuances of meditation experiences that cannot be understood without some practical experience. Having an experiential basis is crucial, but this not to suggest that every clinical professional or researcher of meditation should become avid practitioners of meditation. Instead, learning a common language through meditation practices and even experiencing a range of different meditation techniques are necessary for effectively and clearly communicating the knowledge across scientific and public realms.

Returning to the topic of conducting meditation research, the next step following the definition and description of the targeted meditation practice and technique is determining the outcome variables and how exactly we would measure these outcomes. These steps seem rudimentary, yet methodological concerns precisely begin at these initial steps of experiment planning. Researchers are typically interested in examining the effects of meditation on a specific facet of human functioning. Taking psychological well-being as an example, we may want to study whether meditation could improve psychological well-being (Tang, Tang, & Gross, 2019), but would immediately realize that there are multiple aspects of psychological well-being such as emotions, life satisfaction, and happiness, which means we would need to make a decision about which indexes we should select as outcome variables. Second, after we decide on one or two aspects of psychological well-being that we would like to investigate, the problem of selecting appropriate measurement tools would arise. For emotions, there are numerous self-report questionnaires of positive and negative affect available for research purposes, as well as different objective tasks that could tap into emotions. Additionally, there are even physiological indexes such as skin conductance response that could be used for assessing emotional arousal under various scenarios. Even if we just want to focus on emotion, a major component of psychological well-being, and how meditation could improve emotion, we would still encounter this problem of having too many choices for assessing a single construct. Although this may not be an issue for researchers who have predetermined their choices of assessment, there are still a few methodological issues worth considering before investing in one or two measurement tools. Given the variety of potential choices of measurement, the question is often not simply what to select out of these possibilities, but rather which measurement tool would reliably and accurately assess our interested construct.

Reliability and validity

The selection of assessment tools is often decided based on what had been commonly and widely used in prior literature for assessing the same construct, yet it is important to realize that some of these instruments may have suffered from poor psychometric properties. Psychometrics is a field of study that concerns the theories and techniques for assessing psychological constructs. Its theories have profound implications for constructing and developing instruments and tools for psychological measurement. Notably, psychometric theories have heavily underscored two important concepts regarding various techniques of psychological measurement: reliability and validity. These two terms are definitely no strangers to most researchers, but are not always carefully considered during the planning stage of an experiment. Reliability refers to the overall consistency of a measure, meaning that it should be able to produce similar results after being administered multiple times to the same individual when all other factors are held constant. For example, test—retest reliability is one way of assessing the reliability of a measurement tool. Imagine that we administer a personality questionnaire to a participant at one time point and ask the same person to come back a week later to take the same questionnaire again. The personality scores should, in theory, be highly correlated to one another and should mostly be in agreement and without drastic differences in scores. Therefore a high test—retest reliability would indicate that the instrument is likely to be a reliable tool for assessing personality. The importance of reliability of measurement tools in empirical studies is intuitive, since without a reliable measurement tool, we cannot determine whether changes in our measured outcomes are due to random experimental noises or errors (i.e., unreliable assessment) or are in fact due to our experimental manipulation. Fortunately, nowadays most existing self-report questionnaires have empirical reports focusing specifically on reliability, which should be used to guide the decision making of measurement tools selection. Furthermore, for researchers who conduct cross-cultural investigations, the reliability of translated questionnaires or measurement tools is also important and most instruments have had their translated versions examined for reliability. Relatedly, studies on reliability of cognitive tasks are also gradually emerging, although they are not as common as reliability studies on self-report questionnaires. Overall, as it is the case in other fields of study, reliability should be the primary decision-making factor when researchers choose measurement tools for meditation outcomes.

Validity is another psychometric property critical for ensuring the reproducibility and scientific rigor of our empirical investigations. Validity refers to the extent to which an instrument or assessment tool measures what it intends to measure. For example, validity simply means that if a questionnaire is intended for personality, then its content and questions should be measuring personality, rather than other constructs such as intelligence or life satisfaction. Compared to reliability, validity is generally less problematic or less likely to cause concerns if the construct is clearly defined and operationalized by the measurement tool based on established theories. However, validity could be compromised if there is no clear consensus on the definition and operationalization of a particular construct. This is unfortunately the case for the construct of mindfulness due to the ongoing conceptual ambiguity with regard to the definition. Mindfulness can refer to meditation practices and a dispositional trait. In the case of trait mindfulness, definitions can vary across different self-report questionnaires, yet all of these questionnaires are intended for measuring the construct of mindfulness. In a recent perspective article by Van Dam et al. (2018), this conceptual and semantic ambiguity in defining the construct of mindfulness was extensively discussed. They specifically highlighted that such ambiguity can lead to poor construct validity, since existing self-report questionnaires of mindfulness each have somewhat different semantic definitions of mindfulness. For example, among the nine commonly used mindfulness questionnaires listed by Van Dam et al. (2018), some questionnaires are developed based on Buddhist theory while others are based on cognitive theory, which makes it extremely challenging and somewhat impossible to compare any of the questionnaires. Perhaps the most worrisome problem is that trait mindfulness is also a construct that is commonly used as one of the meditation outcomes. Therefore having measurement tools with various semantic definitions of the same construct would be highly problematic when multiple studies are compared in order to draw general conclusions regarding the effects of meditation on trait mindfulness. This is one of the prominent cases in meditation research where the validity of measurement tools is especially worth attention and careful consideration. However, just like reliability, the issue of validity in meditation research could be addressed by clearly operationalizing definitions of all targeted constructs and selecting instruments that best fit the investigative goals in terms of validity and reliability.

Although emphasizing that reliability and validity of measurement tools are critical for every field of study, not just meditation, these two concepts are worth underscoring for studies adopting an individual differences perspective in empirical investigations. High reliability of a measurement tool would ensure that there would not be erroneous fluctuations of the observed results across different time points of assessment. Hence at the individual level any changes after experimental manipulation could be deemed reliable and stable. Moreover, any individual differences in these changes could also be deemed trustworthy as these changes tap into the true variance following experimental manipulation (i.e. meditation). Additionally, high reliability can ensure that correlational analyses would yield stable relationships between measured variables. This is critical for individual differences investigations, since this type of analysis is common for examining individual differences (i.e., changes in one variable are related to the baseline measure of another variable) and high reliability would ensure consistent conclusions can be made concerning the relationships among these variables. Last, there is another notable phenomenon with respect to reliability and individual differences that is typically found in highly robust cognitive tasks and paradigms. The reliability paradox is where well-established experimental effects can be consistently replicated at the group level using certain cognitive tasks, but not because the tasks are highly reliable, rather it is due to the tasks being less sensitive to between-subject variability (Hedge, Powell, & Sumner, 2018). This paradox is undoubtedly problematic with respect to individual differences investigations since the goal of these studies is to be sensitive to individual variability in outcome variables. Furthermore, data analyses using a correlational approach would also be affected, as these well-validated tasks are not sensitive to between-subject variability and would likely undermine the correlational relationships observed between theoretically important variables, resulting in misleading conclusions. Based on the findings of Hedge et al., (2018) regarding cognitive tasks for attention and cognitive control, robust cognitive paradigms should not be assumed to be highly sensitive to individual differences. In particular, they demonstrated that cognitive tasks that are frequently employed in meditation research (e.g., Stroop, Flankers, and go/no-go tasks) are precisely the ones to have relatively good test−retest reliabilities, typically greater than 0.60 (good/substantial reliability) and less than 0.80 (excellent reliability). However, these tasks do not translate well into correlational studies, precisely due the fact that they showed sensitivity to robust experimental effects at the group level, but cannot clearly distinguish individuals from the group. This kind of paradox is not surprising because many of these tasks were developed decades ago based on the idea that "the goal of the experimentalist is to minimize individual differences" (Hedge et al., 2018). As a result, these tasks may be suitable for making conclusions about general cognitive effects at the group level, but would not be suitable for studies dedicated the investigations of individual differences in these cognitive effects. Thus when the focus of meditation research is on individual differences rather than group differences, then it is imperative for researchers to be highly aware of such problems in the well-established cognitive paradigms and avoid potential pitfalls by exercising caution in study planning and task selection.

Research design

Meditation research typically follow two types of research designs, namely (1) cross-sectional and (2) longitudinal. Cross-sectional studies are often conducted between a group of experienced meditators and a matched group of healthy individuals without any meditation experience. The comparison of any psychological and cognitive measures is made at one time point. Longitudinal studies assess changes within the same group across time, and at least two different time points are involved. Usually a group of novices would participate in a meditation program for a period of time, and changes in outcomes are assessed at baseline and postintervention. The first type of design is particularly useful for examining how long-term meditation experiences may result in lasting impacts on brain and behavior, yet methodological concerns on the inherent individual differences among these experienced meditators are widely discussed (Davidson, 2010; Van Dam et al., 2018). Indeed, with cross-sectional design, it is difficult to ascertain whether or not these observed differences may be attributed to genetic and environmental factors other than meditation experiences, as studies have shown that people with certain personality traits are likely to seek out healthy lifestyle exercises and interventions such as physical training and contemplative practices such as meditation. Therefore researchers using cross-sectional research designs should be mindful of these potential confounds that may contribute to the observed differences across groups. Longitudinal research design also has methodological issues. First, many longitudinal studies lack an adequate control group to which the meditation intervention group could be compared. The reason for having a control group is because it would help identify the contribution of meditation intervention from that of other factors such as group interaction, enthusiasm of the instructors, or placebo effects, which may also lead to the observed changes in brain and behavior (Van Dam et al., 2018). However, the control group should not simply be a waitlist control group, as these usually include participants who do not engage in any types of intervention and would just be assessed twice at the same time as the meditation group. Instead, an active control intervention should be provided to participants in the control group, which would enable rigorous comparisons and address potential contribution resulting from nonspecific factors other than the intervention (Tang et al., 2015). Nevertheless, finding or developing an active control intervention for meditation research could be challenging. There are several suggestions in general (Van Dam et al., 2018) that could be used when selecting and designing an active control intervention: (1) the length and structure of intervention should be matched; (2) the amount of informal practice required outside of formal intervention should be matched; and (3) the expertise and enthusiasm of the instructors should be matched. Researchers should also ensure that the delivery format and intervention type are as closely matched as possible. For instance, if meditation intervention is provided through an online platform, then the control intervention should also follow the same delivery format. Likewise, unless the goal of the study is to compare the effectiveness of two different types of intervention (i.e., physical exercise vs meditation), the interventions should stay similar in terms of the required amount of physical and mental effort. In general, satisfying these basic criteria would greatly improve the scientific rigor of control conditions in meditation research.

Sample size

Another critical issue for individual differences research is the need for a large sample size. The reason for this is intuitive, since without a large sample size, it is impossible to even begin individual differences investigations because there is simply no interindividual variability to be detected. Unfortunately, the sample size in most meditation studies typically fall within the range of 10−50 people per group, with the exception of a few recent large-scale studies that reached more than 50 participants per group (Hildebrandt, McCall, & Singer, 2017; Kuyken et al., 2013). Even for these larger group studies, the sample size of each meditation intervention group was fairly small, but the combined number of multiple groups could build up to a large sample size. The issue of small size in meditation research has been criticized over the years due to the fact that population-level inferences were made based on small sample-size studies (Davidson & Dahl, 2018; Van Dam et al., 2018). Furthermore, generalizations and overinterpretations of the observed benefits are increasingly problematic for scientific research and public engagement in meditation practices. A large sample size is necessary for any type of scientific investigation and is particularly important for individual differences investigations not only because a large sample size is necessary for capturing intersubject variability, but also because a large sample size would enable greater statistical power for detecting effects that are likely to be highly subtle, such as individual variability (Tang et al., 2015; Goldberg et al., 2017; Van Dam et al., 2018).

Potential solutions for individual differences research

Conceptual and methodological issues are not unique to meditation research, but nevertheless are troublesome for studying individual differences in meditation. However, there are few potential remedies with respect to the issue of reliability, validity, and sample size. Regarding reliability, we have highlighted the importance of first examining published reports on reliability of the measurement tools prior to the final selection. The bottom line is that measurement tools with low reliability should not be selected and, more importantly, it is imperative to avoid task paradigms that may have relatively good reliability, but are known to have low sensitivity to interindividual variability (see Hedge et al., 2018). This reliability paradox is especially prominent for cognitive tasks, but less so for self-report questionnaires, which are generally aimed for capturing individual differences in the constructs. Aside from making informed decisions during experiment planning, after data are acquired, additional steps should be taken to address the issue of reliability. In particular, it is necessary to compute and report reliability coefficients of the measured results in research findings since assessing reliability can serve as a crucial quality-control step, allowing the researchers to qualitatively evaluate the extent to which the observed effects can be trusted and adding a further step of validation when interpreting and reporting study results. For validity, the targeted construct should be as clearly defined and operationalized as possible by the measurement tool and special caution should be exercised if the construct has multiple potential semantic meanings, such as the case with trait mindfulness. It is important to note that a measurement tool may be reliable, but is not necessarily valid, thus, ensuring validity would rely largely on the judgment of the researchers.

For the issue of sample size, it is necessary to first acknowledge that acquiring data with large sample size is not easily accomplished for research studies that involve psychological interventions. There are several substantial constraints for conducting longitudinal studies with psychological interventions. First, it is suboptimal to have a large intervention group with more than 20 people as the quality of intervention may be jeopardized due to the instructor being unable to attend to everyone. During the intervention program, the instructor needs to be able to provide instructions, respond to questions, and monitor the progress of each participant. For this reason, previous studies have mostly adopted a sample size of under 30 people per group for meditation intervention. This implies that acquiring a large sample size would take more time and resources than typical research studies, in order to accumulate a large number of participants through multiple intervention groups. Currently there is no better way to speed up the process than by running multiple intervention groups in parallel. However, there is also another alternative to potentially provide intervention to a large group of people, that is, developing online intervention programs. Mobile applications that include meditation practices have been fairly popular among the general public, but the extent to which these applications may be effective in improving outcomes has not been explored. Likewise, web-based meditation programs have been developed over recent years, but the standardization and effectiveness of such programs warrants further investigation. A recent review suggests some support for the effectiveness of web-based meditation programs in improving psychological well-being, with most studies showing large effect sizes (Fish, Brimson, & Lynch, 2016). However, these studies typically do not have a control group to which the meditation effects may be compared, making it impossible to conclude that such web-based programs are indeed effective for improving psychological outcomes. Another issue in longitudinal studies is participant attrition, which also unfortunately prolongs data collection for a large sample size. This challenge may be potentially addressed by offering greater incentives for study completion and increasing flexibility to accommodate the needs of study participants.

Measuring meditation outcomes

We discussed the major conceptual and methodological issues in conducting meditation research as well as challenges specifically pertaining to individual differences investigations. We provided some potential solutions to methodological constraints typically found in longitudinal studies with psychological interventions. In this section we will briefly discuss measurement tools that would be useful for assessing meditation outcomes. However, our goal is not to provide a list of tasks or questionnaires; rather we would like to broadly survey the tools available for meditation research and discuss novel techniques that could be useful in providing additional insights into our existing knowledge of meditation effects on psychological well-being and cognition.

Self-report questionnaires

Self-report is one of the most direct options for inquiring about psychological well-being from a first-person perspective. Although the subjective nature of self-report questionnaires is not without criticism, their convenience and utility in tapping into psychological health have been validated. For clinical diagnoses of psychiatric disorders, interviews and questionnaires are always used for evaluating the psychological state of the patient. In general, well-validated and reliable questionnaires are highly useful if the targeted outcomes are related to different aspects of psychological well-being. For the clinical population, questionnaires assessing psychological symptoms are readily available for researchers who would like to investigate the effects of meditation on improving psychological health. Similarly, for the healthy population, the same questionnaires could also be effective in measuring subclinical symptoms, and whether or not meditation would be beneficial in enhancing these aspects of psychological health. However, self-report questionnaires are unlikely to be accurate for cognitive-related constructs such as self-report capacity of attention control and cognitive control. People usually tend to either underestimate or overestimate their ability as these self-evaluations are not as straightforward as expressing their opinions or how they feel emotionally. For reporting practice time and experience, self-report questionnaires are also the most straightforward instruments as there are no other ways to noninvasively tapping into the amount of practice an individual engages in. However, these self-report data should be treated with caution since these reports are, again, estimations of past experiences, which would require reflecting from memory, especially for those with long-term meditation experiences. Another caveat of self-report meditation practice is that the quality and quantity of meditation practices outside of formal intervention settings are difficult to control and are subject to unforeseen noises that are unlikely to occur in formal practices. However, experience sampling methods could potentially be promising for collecting data on informal practices, as this kind of method would prompt participants from time to time through mobile devices to inquire about their experiences using simple questions that allow researchers to collect immediate/real-time responses. This experience sampling approach can be particularly useful for meditation research and can collect data on both quantity and quality with respect to meditation practices.

Cognitive tasks

Assessing cognitive changes following meditation intervention remains one of the most challenging endeavors. The reason for this is that cognitive abilities involve complex mental representations and processes and we are still in the infancy stage of understanding the intricacy of human cognition and developing tools to better measure different dimensions of cognitive function. Furthermore, given the reliability issue with regard to cognitive paradigms, measuring cognitive changes are even more challenging as we need to be mindful of tasks that are insensitive to individual variability. Another difficulty of assessing cognitive function in meditation research arises from the wide array of cognitive tasks that has been used by researchers. Often there are multiple tasks available for assessing a single cognitive construct (e.g., working memory). Therefore it is impossible to make constructive and objective recommendations on which tasks to use for which cognitive construct in meditation research. Perhaps a better approach for reconciling the choice of cognitive tasks is to utilize a standardized cognitive task battery with well-validated and reliable paradigms that can tap into essential components of human cognitive function and capacity. Fortunately, such a cognitive task battery exists. Recently, the National Institutes of Health (NIH) toolbox (http://www.nihtoolbox.org) has been advertised by the National Institute of Health as a comprehensive set of measurements that can quickly assess cognitive, emotional, motor, and sensory processes. The toolbox was developed as part of the Blueprint-funded NIH Toolbox for Assessment of Neurological and Behavioral function and is designed to be used in healthy individuals between the ages of 3 and 85 years (Barch et al., 2013). A very appealing feature of this NIH task battery is that all tasks were developed and validated to be suitable for computerization, meaning that these tasks could simply be administered through mobile devices. This greatly speeds up the process of data collection and analyses, since data entry is likely to be minimized and data analyses are also likely to be accelerated given that everything is computerized. More importantly, using the NIH toolbox could allow researchers to compare their data with a nationwide representative sample, having a basic idea of where their study samples fall within the large population sample. In fact, the NIH-funded Human Connectome Project involving a sample size of 1200 participants have utilized the NIH toolbox as the primary task battery for assessing cognitive function and other related processes. Last, the NIH toolbox was created to be suitable for longitudinal studies, which is another desirable feature for meditation research that often involve longitudinal designs with multiple time points of assessment.

For cognitive function, the NIH toolbox currently includes the following domains: (1) episodic memory (picture sequence memory); (2) executive function/cognitive flexibility (dimensional change card sort); (3) executive function/inhibition (Flanker task); (4) language/vocabulary comprehension (picture vocabulary); (5) processing speed (pattern completion processing speed); (6) working memory (list sorting); and (7) language/reading decoding (oral reading recognition) (Gershon et al., 2013; Zelazo et al., 2013). Although the NIH toolbox may not include all domains of cognition, the strengths and advantages of this standardized battery are worth exploiting in future investigations. Furthermore, amid the problematic selection of appropriate and reliable cognitive assessments for meditation research, employing the NIH toolbox is also likely to mitigate these common methodological concerns and improve the rigor of meditation research.

Neuroimaging

We discussed neuroimaging studies of meditation, including detailed description of commonly used techniques [e.g., functional magnetic resonance imaging (fMRI), structural magnetic resonance imaging] in Chapter 6, Brain regions and networks in meditation, thus, we will not reiterate each of the techniques here. However, we will highlight some of the promising analytical approaches relevant for individual differences investigations of meditation. For functional neuroimaging, both task fMRI and resting-state fMRI have been shown to have good sensitivity in capturing individual differences in brain activation and connectivity (Dubois & Adolphs, 2016; Gratton et al., 2018). Although neuroimaging techniques are widely applied in research studies of meditation, the question is which indexes should we be using to explore potential interindividual variability, given that there are various and sophisticated analytical techniques at our disposal. Based on existing literature, the most straightforward approach would be to examine functional connectivity among brain regions. The fingerprinting approach demonstrated by Finn et al., (2015) reliably and robustly showed that individual resting-state functional connectivity profiles can act like fingerprints to distinguish an individual from the rest of the group. This may hold immense promise for individual differences investigation as we know that individuals are more similar to themselves in terms of functional connectivity patterns than anyone else. For meditation research, exploring changes in whole-brain functional connectivity profiles across individual participants may be able to capture individual-specific changes related to meditation. Moreover, these functional connectivity profiles can predict cognitive behavior such as fluid intelligence, which is a critical marker for human cognitive capacity and is also related to different life outcomes (Finn et al., 2015). Therefore functional connectivity profiles could be useful markers for examining individual differences in brain and behavioral changes following meditation. Interestingly, many of the more complicated functional neuroimaging analyses are derived from basic functional connectivity analyses. For example, we briefly discussed the utility of examining meditation states and effects at the level of brain networks rather than at isolated regions. Brain networks have recently been investigated using graph theoretical analyses that require functional connectivity data as input for computing brain network properties and indexes. Graph theoretical analyses treat the brain as a complex system, where brain regions function as nodes and are connected either structurally through white matter or functionally via temporal correlations. A growing body of literature has already investigated brain network organization using graph theoretical analyses and have identified key network hubs within the brain that serve critical functions in supporting cognition and behavior (Bullmore & Sporns, 2009, 2012). More importantly, these network measures have shown potential in illustrating individual differences in brain organization as well as in predicting behavior (Tompson, Falk, Vettel, & Bassett, 2018).

Finally, another exciting approach in neuroimaging analysis is the multi-voxel pattern analysis (MVPA) or sometimes referred to as machine learning. The idea of MVPA is to train a classifier to detect distributed patterns of brain activation during various mental states and representations and apply this classifier in the decoding of mental states. MVPA is sensitive to distributed patterns of brain activity, which distinguishes it from commonly used mass-univariate analyses that compare mean differences among different conditions and mental states. The strength of MVPA is that it considers the contribution from individual brain voxels rather than focuses on the averages of multiple voxels, which tend to blur out fine-grained information relevant for mental processes and states (Norman, Polyn, Detre, & Haxby, 2006; Pereira, Mitchell, & Botvinick, 2009). Meditation research would particularly benefit from MVPA since potential similarities and differences of various meditation techniques at the level of the brain may be more easily captured by MVPA, rather than using the mass-univariate approach that compares the mean activation between techniques. Furthermore, the meditative state can be considered as a form of mental states, which means its neural representations are likely to behave in a state-like manner through distributed patterns of brain activation. Hence exploiting the advantages of MVPA in neuroimaging studies of meditation would likely reveal novel insights into the underlying brain mechanisms of meditation. For example, our series of randomized studies have shown that one form of mindfulness meditation, integrative body–mind training (IBMT), can improve attention, emotion regulation, and cognitive performance through changing brain activity and structural connectivity. To examine whether and how the short-term mindfulness meditation practice alters large-scale brain networks, we applied MVPA to resting-state fMRI (rsfMRI) data to identify changes in brain activity patterns and assess the neural mechanisms induced by brief training. Whole brain rsfMRI was performed on a college student group who received 2 weeks of IBMT with 30 min per session (5 h training in total). Classifiers were trained on measures of functional connectivity in the fMRI data and they were able to reliably differentiate (with 72% accuracy) patterns of connectivity from before versus after the IBMT training. After training, an increase in positive functional connections (60 connections) were detected, primarily involving the bilateral superior/middle occipital gyrus, bilateral frontal operculum, bilateral superior temporal gyrus, right superior temporal pole, bilateral insula, caudate, and cerebellum. These results suggest that brief mental training alters the functional connectivity of large-scale brain networks at rest that may involve a portion of the neural circuitry supporting attention, cognitive and affective processing, awareness, sensory integration, and reward processing (Tang, Tang, Tang, & Lewis-Peacock, 2017).

It is also possible to use MVPA to classify between experienced meditators and novice meditators based on patterns of brain activation, which may be able to provide information related to the brain basis of meditation expertise. Taken together, for any meditation studies that seek to employ neuroimaging, bridging analyses of neuroimaging data and behavioral data would ultimately be the most informative approaches for uncovering the practical implications of any meditation-related changes. Without relating the brain to behavior, the interpretations with respect to behavior are, at best, speculative.

Summary and implications for translational research

We discussed the issues of reliability, validity, and sample size with regard to meditation research. These issues are also found and emphasized in other fields of scientific investigations, which suggests an increasing universal awareness of these conceptual and methodological issues within the scientific community. For future investigations, being mindful of these critical issues would greatly improve our methods of assessments and experimental designs, thereby enhancing the scientific rigor and reproducibility of our empirical findings. However, there are still outstanding issues regarding meditation research that have yet to be resolved, namely the issues of defining and measuring the construct of mindfulness, as well as characterizing meditation practices and techniques into meaningful categories for more accurate interpretation of results. We also described promising measurement tools and analytical approaches sensitive to individual differences, which could be useful in capturing meditation-related changes as well as individual differences in these changes. Overall, the complexity and heterogeneity of meditation effects on the brain and behavior call for further thorough investigation by incorporating more rigorous means of research design, assessment, and analyses. Delineating the precise underlying mechanisms of meditation by utilizing interdisciplinary methodologies and knowledge is imperative for scientific understanding and applications of meditation.

References

Barch, D. M., Burgess, G. C., Harms, M. P., Petersen, S. E., Schlaggar, B. L., Corbetta, M., ... Nolan, D. (2013). Function in the human connectome: Task-fMRI and individual differences in behavior. *Neuroimage, 80*, 169–189.
Bullmore, E., & Sporns, O. (2009). Complex brain networks: Graph theoretical analysis of structural and functional systems. *Nature Reviews Neuroscience, 10*(3), 186.
Bullmore, E., & Sporns, O. (2012). The economy of brain network organization. *Nature Reviews Neuroscience, 13*(5), 336.

Davidson, R. J. (2010). Empirical explorations of mindfulness: Conceptual and methodological conundrums. *Emotion, 10*(1), 8–11.

Davidson, R. J., & Dahl, C. J. (2018). Outstanding challenges in scientific research on mindfulness and meditation. *Perspectives on Psychological Science, 13*(1), 62–65.

Davidson, R. J., & Kaszniak, A. W. (2015). Conceptual and methodological issues in research on mindfulness and meditation. *American Psychologist, 70*(7), 581.

Dubois, J., & Adolphs, R. (2016). Building a science of individual differences from fMRI. *Trends in Cognitive Sciences, 20*(6), 425–443.

Finn, E. S., Shen, X., Scheinost, D., Rosenberg, M. D., Huang, J., Chun, M. M., . . . Constable, R. T. (2015). Functional connectome fingerprinting: Identifying individuals using patterns of brain connectivity. *Nature Neuroscience, 18*(11), 1664.

Fish, J., Brimson, J., & Lynch, S. (2016). Mindfulness interventions delivered by technology without facilitator involvement: What research exists and what are the clinical outcomes? *Mindfulness, 7*(5), 1011–1023.

Gershon, R. C., Wagster, M. V., Hendrie, H. C., Fox, N. A., Cook, K. F., & Nowinski, C. J. (2013). NIH toolbox for assessment of neurological and behavioral function. *Neurology, 80* (11 Supplement 3), S2–S6.

Goldberg, S. B., Tucker, R. P., Greene, P. A., Simpson, T. L., Kearney, D. J., & Davidson, R. J. (2017). Is mindfulness research methodology improving over time? A systematic review. *PLoS One, 12*(10), e0187298.

Gratton, C., Laumann, T. O., Nielsen, A. N., Greene, D. J., Gordon, E. M., Gilmore, A. W., . . . Dosenbach, N. U. (2018). Functional brain networks are dominated by stable group and individual factors, not cognitive or daily variation. *Neuron, 98*(2), 439–452.

Hedge, C., Powell, G., & Sumner, P. (2018). The reliability paradox: Why robust cognitive tasks do not produce reliable individual differences. *Behavior Research Methods, 50*(3), 1166–1186.

Hildebrandt, L. K., McCall, C., & Singer, T. (2017). Differential effects of attention-, compassion-, and socio-cognitively based mental practices on self-reports of mindfulness and compassion. *Mindfulness, 8*(6), 1488–1512.

Kuyken, W., Weare, K., Ukoumunne, O. C., Vicary, R., Motton, N., Burnett, R., . . . Huppert, F. (2013). Effectiveness of the Mindfulness in Schools Programme: Non-randomised controlled feasibility study. *The British Journal of Psychiatry, 203*(2), 126–131.

Norman, K. A., Polyn, S. M., Detre, G. J., & Haxby, J. V. (2006). Beyond mind-reading: Multivoxel pattern analysis of fMRI data. *Trends in Cognitive Sciences, 10*(9), 424–430.

Pereira, F., Mitchell, T., & Botvinick, M. (2009). Machine learning classifiers and fMRI: A tutorial overview. *Neuroimage, 45*(1), S199–S209.

Tang, Y. Y. (2017). *The neuroscience of mindfulness meditation: How the body and mind work together to change our behavior*. Springer.

Tang, Y. Y., Tang, R., & Gross, J. (2019). Promoting psychological well-being through an evidence-based mindfulness training program. *Frontiers in human neuroscience, 13*, 237.

Tang, Y. Y., Hölzel, B. K., & Posner, M. I. (2015). The neuroscience of mindfulness meditation. *Nature Reviews Neuroscience, 16*(4), 213.

Tang, Y. Y., Jiang, C., & Tang, R. (2017). How mind-body practice works - integration or separation? *Frontiers in Psychology, 8*, 6.

Tang, Y. Y., Tang, Y., Tang, R., & Lewis-Peacock, J. A. (2017). Brief mental training reorganizes large-scale brain networks. *Frontiers in Systems Neuroscience, 11*, 6.

Tompson, S. H., Falk, E. B., Vettel, J. M., & Bassett, D. S. (2018). Network approaches to understand individual differences in brain connectivity: Opportunities for personality neuroscience. *Personality Neuroscience, 1*. Available from https://doi.org/10.1017/pen.2018.4.

Van Dam, N. T., van Vugt, M. K., Vago, D. R., Schmalzl, L., Saron, C. D., Olendzki, A., ... Fox, K. C. (2018). Mind the hype: A critical evaluation and prescriptive agenda for research on mindfulness and meditation. *Perspectives on Psychological Science, 13*(1), 36−61.

Zelazo, P. D., Anderson, J. E., Richler, J., Wallner-Allen, K., Beaumont, J. L., & Weintraub, S. (2013). II. NIH Toolbox Cognition Battery (CB): Measuring executive function and attention. *Monographs of the Society for Research in Child Development, 78*(4), 16−33.

Chapter 9

Personalized meditation

Although we often find and get used to people who use the same brand of smartphones, wear the same kind of T-shirts, or participate in the same training course as we do, the idea of personalizing these things and experiences based on our needs and preferences is extremely appealing. The cost of personalization is not trivial, which is also why the idea does not really come across our mind when we think about the things or services that could have been personalized according to our needs. Personalization in daily life may be as important as it is in health care, since in health care, our physical and psychological conditions are often unique to ourselves, suggesting that personalizing health care and its services may be especially important for ensuring the quality and efficacy of treatment. Typically, we have become accustomed to being prescribed common drugs or therapies that have been offered to most people and rarely think about how we may react differently to these medications and therapies. However, if individual differences are constantly distinguishing us from rest of the group, then considering more personalized approaches to health care is necessary for health improvements at the individual level. The concept of *personalized medicine* has been around for some years and was recently put forward and heavily emphasized by leaders from governmental agencies. The idea advocates for the development of personalized treatment plans and methods based on biological markers, notably genes, since genetic variations among individuals have repeatedly been implicated in human illnesses (Hamburg & Collins, 2010). It is important to take note of the subtle distinction between precision medicine and personalized medicine—the former refers to identifying which treatment would be best for which individual, whereas the latter refers to developing personalized treatment approaches for each individual. Hence personalized medicine takes a step further to design and tailor treatment approaches specifically for each individual. However, both concepts overlap in many ways as their ultimate goals are all to enhance the overall experience and effectiveness of treatment for individual patients. Furthermore, both approaches require a careful evaluation and consideration of individual characteristics, traits, biological variations, and other relevant factors in order to make an informed decision with regard to choosing and developing appropriate and effective treatment approaches for each individual. Sometimes these two

The Neuroscience of Meditation. DOI: https://doi.org/10.1016/B978-0-12-818266-6.00010-1

terms are used interchangeably, but readers should take note of the subtle differences between the two approaches.

Personalized medicine can obviously be extended beyond the medical domain, which primarily focuses on treating diseases such as cancer and other serious physical illnesses through personalized biological therapies that use human embryonic stem cells from the individuals to achieve targeted outcomes (Hamburg & Collins, 2010). In the psychological realm, the philosophy of personalized medicine can also be critical for improving the effectiveness of psychotherapy and interventions, although we would likely be using different markers and technology than those used in the medical realm. Individual specific factors other than genetic variations could also be the center of the focus when we aim to develop personalized psychological intervention approaches. Interestingly, the notion of personalizing psychological therapies and interventions has already been theorized and put into action by some exciting research. A field of study focusing on potential genetic determinants of response to psychological interventions and therapies, namely *therapygenetics*, has gained considerable research progress and support in children diagnosed with anxiety disorders (Beevers & McGeary, 2012; Lester et al., 2012). Specifically, one study in children diagnosed with anxiety disorders found that genetic variations in the promoter region of the serotonin transporter gene (5-HTTLPR) predicted their responsiveness to cognitive behavior therapy (CBT) (Eley et al., 2012). The study found that children who carry two copies of the short allele had better treatment response at 6-month follow-up than children who carry one or two copies of the long allele. Another study based on therapygenetics showed that children diagnosed with anxiety disorders and who received CBT were more likely to be free of primary anxiety response at follow-up if they had one or more copies of the T allele of the nerve growth factor (NGF) (rs6330) gene (Lester et al., 2012). However, in another study, the association between 5HTTLPR genotype and CBT outcome did not replicate. Short-allele homozygotes showed more positive treatment outcomes, but with small, nonsignificant effects. Future studies would benefit from utilising whole genome approaches and large, homogenous samples (Lester et al., 2016). CBT is a close analog of meditation-based interventions, thus these findings of genetic polymorphisms related to responsiveness to psychological interventions may also be relevant for predicting the responsiveness of meditation-based interventions in children diagnosed with anxiety disorders. Although it is unclear whether these genetic markers would apply to different populations such as adults with anxiety and other psychiatric disorders, these findings do suggest the promising potential of using genetic markers in psychotherapies and interventions to inform treatment planning and selection at the individual level. More importantly, knowledge gained from therapygenetics may be instrumental for moving the field of psychological interventions toward the personalization of therapies and interventions (Beevers & McGeary, 2012).

These results are consistent with findings in genetic and epigenetic research discuss in Chapter 4, Genetic association with meditation learning and practice (outcomes).

In light of these preliminary, yet promising, lines of evidence with respect to personalizing psychological interventions, we propose the concept of personalized meditation. If we are able to use genetic markers for predicting individual responsiveness to psychological intervention, just like using genetic markers to guide individual-based medical treatment approaches, then it is also possible and highly feasible to explore the potential of designing personalized approaches to different types of psychological interventions such CBT and meditation. Complementary and alternative medicine (CAM) approaches have been discussed with respect to their potential in personalized health care, since the popularity of CAM interventions such as meditation and yoga has attracted more individuals in recent decades than ever before (Edwards, 2012). Based on national surveys sponsored by the National Center for Complementary and Integrative Health, many patients begin to seek out CAM interventions in addition to mainstream medicine, consequently putting CAM interventions into the spotlight of health care along with conventional treatment approaches. It is therefore appropriate to initiate the discussion of personalized meditation, as meditation is not only an important form of CAM interventions, but is also a major type of psychological intervention used by clinicians to treat certain psychiatric disorders. As our understanding of the underlying mechanisms and effects of meditation-based intervention grow, individual differences in response to this intervention have also gained more traction from the scientific community, which is likely to contribute to the development of more personalized approaches to improve health and well-being. To accomplish this exciting and challenging endeavor of personalizing meditation-based interventions, we need to consider two essential components of enabling personalized strategies in psychological intervention—prediction and implementation.

Prediction

If we cannot make basic predictions about what kind of intervention approaches may be suitable for each individual, then we cannot even begin the process of designing personalized interventions. Hence the initial stage of developing and planning any personalized interventions would be to utilize all relevant individual-specific information to generate a basic profile of predictions for each individual. For example, when an individual comes into see a healthcare professional (i.e., clinicians, psychiatrists) about his or her depression, the healthcare professional has a wide range of treatment approaches available for treating depression, but which intervention(s) should the professional select and what kind of personalization does the professional need to make in order to maximize treatment effectiveness for this individual? From a personalized medicine perspective, the professional would first need to evaluate and understand preferences, characteristics, biological, and psychological conditions of the individual and gather as much information as possible to make informed decisions about their treatment plan. Back to where we started our discussion of individual differences in response to psychological interventions, there are several dimensions of individual differences that could be used to make predictions about individual responsiveness. Of note, we are aware of the lack of sufficient empirical findings to support the predictive power of specific individual characteristics or factors for intervention responsiveness, but would like to highlight some promising domains to facilitate further research and discussion of personalizing psychological interventions, particularly meditation.

Genetic variations in the polymorphisms of different genes have shown promise in predicting individual variability in response to psychological interventions. Although genotyping is uncommon in the field of psychotherapy and related interventions, the utility of this technology is demonstrated by therapygenetics (Beevers & McGeary, 2012). The cost of genotyping has decreased considerably over the past decade, and more commercial companies (e.g., 23andMe) are offering affordable genotyping services to the general public and has joined scientists in the effort to understand the genetic contribution to behavior, health, and diseases. Given this convenient accessibility of genotyping services, it is not difficult to incorporate genotyping as part of the regular diagnosis for psychological disorders. However, the challenges would arise after we genotyped the individual—which genes and polymorphisms should we be looking for? The answer would largely depend on scientific knowledge of candidate genes relevant for the individual's disorder as well as for intervention responsiveness, yet both require further research. This is perhaps the most challenging hurdle in personalized medicine for medical treatment approaches and psychological interventions since identifying candidate genes would require considerable scientific progress in uncovering these biological targets. However, at least for children with anxiety disorders, some evidence suggested that the 5-HTTLPR and NGF rs6330 genes may be promising predictors of responsiveness to CBT. Continuing with our example, assuming that we have candidate genes from which we can make predictions about an individual's responsiveness to CBT and meditation. The next step would be to also include other individual differences factors into making our predictions. For genetic markers, the extent of their effects on intervention responsiveness are usually modest (Lester et al., 2012), meaning that there are other contributors to differential intervention responsiveness. Therefore when we make predictions of individual responsiveness to interventions, our model of predictions should not exclusively involve only one or two dimensions of individual variability.

In Chapter 2, Personality and meditation, we described how each of the Big Five dimensions of personality could influence individual responsiveness to psychological interventions. Comparing to our knowledge on the role of genetic markers in intervention responsiveness, we have a slightly better understanding of the potential impacts of personality traits on psychological interventions. Generally speaking, the assessment of personality traits can be easily administered with clinical interview. For our example, the individual can complete personality questionnaires along with other instruments inquiring about psychological health and well-being. Based on the current understanding of the role of personality traits, all five dimensions are likely to be useful to include in the prediction model. We will not reiterate here how each of these traits is relevant for differential intervention responsiveness as this discussion can be found in Chapter 2, Personality and meditation. However, we would like to briefly discuss how personality traits are especially important for personalized medicine and preventive medicine. From the macrolevel, personality traits are the most obvious features of individual differences, likely to be the results of both genetic and environmental influences (with 40%−50% heritability based on twin studies) and can also be considered as phenotypic dimensions of human variability given their relatively stable nature (Chapman, Roberts, & Duberstein, 2011). Additionally, personality traits also play an important role in predicting life outcomes, psychological health, and even longevity. In particular, research on personality psychology has suggested that a higher level of conscientiousness is associated with greater longevity, consistent with the findings of conscientiousness correlating positively with psychological health (Chapman et al., 2011). Although we would obviously want to use personality traits in predicting intervention responsiveness when individuals seek treatment for their problems, it is important to also realize that personality traits hold immense promise for screening individuals in need for preventive interventions. For instance, if an individual is at risk for developing depression and anxiety based on his or her personality profile (e.g., high level of neuroticism), then we may want to personalize preventive strategies in advance before he or she even develops depression. Personality may complement genetic information to increase the predictive power of our model on intervention effectiveness as these two dimensions of human variability are largely intercorrelated (Chapman et al., 2011).

There are also preexisting individual differences in cognitive abilities and psychological well-being that are worth taking into account when predicting individual variability in response to psychological interventions. For individuals who seek psychological treatment, their psychological well-being is likely to be compromised from the beginning, which means that the range of individual variability in the level of psychological health is going to be restricted to the lower end of the spectrum. Mixed evidence has been found with regard to intervention responsiveness of individuals with low levels of psychological well-being as measured by symptom severity. Preliminary evidence seems to suggest that those with very severe psychological symptoms are likely to benefit more from pharmacotherapeutic approaches than from psychological interventions (see more discussion on this in Chapter 2: Personality and meditation). If an individual shows a considerable degree of dysfunction, he or she may not be able to engage in the mental activities required by psychological interventions. However, for individuals exhibiting subclinical symptoms, there does not seem to be a clear trend with respect to which kind of treatment approach is more suitable. Therefore even though symptom severity should be considered in the prediction model when health-care professionals select medications or psychological interventions for individual patients, careful consideration and discussion with the patients are necessary before deciding on either of the approaches.

Individual differences in cognitive abilities could also play a role in affecting intervention responsiveness. Taking attention control as an example, we know that it is one of the core processes involved in meditation and is fundamental to the cultivation of meditative states since successful control of attention is imperative for noticing mind-wandering and directing attentional focus back to the target of concentration (Tang, Hölzel, & Posner, 2015). We would expect individuals with high levels of attention control to have an easier time practicing meditation since they tend to be more skillful in their ability to direct and maintain attention than those with low levels of attention control. There is also evidence suggesting that attention control is positively correlated with trait mindfulness, the very capacity meditation seeks to promote, further suggesting that better attention control capacity can facilitate meditation practices (Walsh, Balint, Smolira, Fredericksen, & Madsen, 2009). These convergent findings imply that preexisting individual differences in abilities necessary for meditation practices could potentially impact how well an individual practices and responds to meditation. Differences in general fluid intelligence may also affect intervention effectiveness through influencing the ability to learn. General fluid intelligence is critical for successful and effective learning and consists of basic abilities such as problem solving and reasoning that are important for facilitating the acquisition of new skills and knowledge (Conway, Kane, & Engle, 2003). For individuals who just begin to learn and practice meditation, having a high baseline level of fluid intelligence would likely lead to greater efficiency in learning new skills from the interventions and in applying such skills in everyday life, which, in turn, would likely lead to better intervention effectiveness in improving targeted outcomes. However, it should be noted that there are definitely other preexisting differences in aspects of cognitive function related to intervention responsiveness. Attention control and fluid intelligence are just two examples for illustrating the utility of considering inherent differences in cognitive abilities. Personalizing intervention approaches based on preexisting individual differences in cognitive abilities and psychological well-being, as well as genetic variations and personality traits, could potentially be very powerful. However, we should be careful about committing to a specific direction of each of these factors since we do not have enough information about how these factors may interact to influence the extent of intervention responsiveness. Moreover, we still need further empirical support from translational research to determine the utility of these individual-specific factors and how such information can be best leveraged by personalized medicine.

Making predictions is crucial for guiding subsequent personalization of intervention approaches, yet individual preferences and values may sometimes be at odds with our suggested approaches. For instance, we may encounter cases in which the patient would like to take medication for treating his or her psychological symptoms, even though his or her symptom severity may be better suited for psychological interventions with less side effects. In this kind of scenario, respecting individual preferences and decisions should always be the top priority as the goal of personalized medicine is all about tailoring the treatment approaches to the needs and values of the individual. Therefore if an individual has strong opinions about his or her treatment plans, valuing the individuality of the patient should always come before any of our predictions or recommendations.

Implementation

After we have a relatively good understanding of the kinds of intervention that are likely to be effective for the individual, the next important step is to design and implement the personalized treatment approaches. There are two possibilities: (1) selecting existing intervention programs based on individual needs and characteristics; and (2) developing and improvising existing intervention programs into a personalized intervention. The former strategy is almost equivalent to the idea of precision medicine, but still relies on discussing and assessing individual characteristics and factors to come up with the best treatment plan for each individual. The latter strategy is likely to involve more effort and work since some necessary improvements and adaptations would need to be made to accommodate the needs of different individuals. We will focus our discussion on developing personalized intervention approaches with a special emphasis on how to personalize meditation practices as this area has largely remained unexplored and could pose considerable challenges with regard to the feasibility of implementing such approaches.

Meditation-based interventions are easier to be personalized than other forms of psychological intervention since there is a wide range of meditation techniques to choose from and adapt. Let us take the standardized mindfulness-based stress reduction (MBSR) program as an example to discuss how we can personalize the program according to individual needs. Within the MBSR program, except for other methods such as Yoga, group support, there are at least four major meditation techniques that are taught to participants, including focused attention, open monitoring, loving kindness, and body scan. Focused attention involves directing and maintaining our attention to a chosen target of concentration, which is usually our breathing pattern (i.e., each inhale and exhale). Open monitoring does not involve a specific target of focus, rather, anything that arises in the present moment is the object of focus, which could thoughts, emotions, or sensations. Loving kindness involves fostering a sense of love, warmth, and friendliness toward oneself and others, which is usually done with imagery or words. Finally, body scan is a technique that involves a moving attention focus to different parts of our body and is usually guided by instructors or the participants themselves. The moving attention focus usually starts from the forehead and proceeds to other parts of the face and then moving down to different body parts. The idea is to observe any sensations that arise from these body parts during the practice. If we would like to personalize MBSR for individual participants, then we need to first determine which of the four techniques would be most effective for treating the targeted symptoms. Given that MBSR is often provided as a multifaceted package to all participants (Davidson and Kabat-Zinn, 2004; Smith, 2004), personalizing this program would entail selecting the most useful strategies based on individual needs and preferences. As there are individual preferences for specific meditation techniques, thus, in addition to selecting techniques for treating targeted symptoms, individual preferences would also need to be assessed.

Assessing individual needs and preferences

Using questionnaires, we can first obtain a basic understanding of individual preferences. Some of the questions may include preferences for intervention settings (e.g., group-based and individual-based) and format (e.g., online and face-to-face). Second, it is also useful to briefly expose participants to each of these techniques and let them practice a few times to determine their preferences. If they find only one or two techniques to be extremely useful and easy to grasp, then we would have concrete information to tailor the meditation intervention based on their preferences. Indeed, a recent study by Anderson and Farb (2018) illustrated that there are individual preferences for using specific modalities of sensory anchors for maintaining attention focus during meditation practices, which further corroborates the notion of assessing individual preferences at the beginning of the intervention before prescribing individuals to a multifaceted intervention package. Equipped with this individual-specific information, we can potentially tease apart the MBSR program and adapt it to fit the needs of the individual through formatting the intervention with the preferred techniques and modified the standardized curriculum with instructions and practices dedicated to these techniques. Moreover, it is also possible to personalize meditation-based intervention based on needs for both healthy and clinical populations. A recent attempt to objectively determine the needs of individuals based on electroencephalogram (EEG) assessments was examined in healthy individuals (Fingelkurts, Fingelkurts, & Kallio-Tamminen, 2015). The study tested the utility of quantitative electroencephalogram (qEEG) for evaluating cognitive and psychological functions of individuals. By comparing the individual's qEEG measures with that of a normative database, the researchers generated individual profiles for each participant that described areas of cognitive and psychological functions that need further improvement. Following this preliminary assessment, the researchers tailored the meditation practices to specifically improve these domains of functions and found this personalized meditation approach induced better outcomes in these areas than a nonpersonalized meditation approach. It is, however, unclear how exactly the researchers tailored and personalized the meditation practices, but the notion of using baseline physiological assessment (e.g., EEG, heart-rate variability, and skin-conductance response) to guide the personalization of meditation-based intervention holds promise for future applications (see Chapter 5: Sympathetic and parasympathetic systems in meditation). The same is true for using genomic information for personalizing meditation practices. For instance therapygenetics, although it has yet to be investigated and examined in meditation research, its implications for clinical applications could be profound in terms of intervention effectiveness and cost-effectiveness.

Personalized meditation and resources

One challenge to the implementation of personalized meditation would be resources. We know that meditation-based intervention is typically offered in group settings where an instructor teaches and guides participants on how to practice different meditation techniques. If we start to personalize meditation for individual participants, then putting them into the same group would not be feasible for this approach. Therefore more effective planning and use of resources would need to be enacted for successful implementation of personalized meditation. For instance, it is possible to have multiple instructors holding different sessions for teaching each technique, which would allow individuals to flexibly attend the sessions they need. Establishing online platforms for holding regular courses to teach different meditation techniques is another alternative, which would likely increase the number of participants per session, since more individuals would be able to attend the courses without being physically present at the scene of action. Indeed, online courses are becoming more feasible and popular these days and many universities have started to offer regular online courses with formal periodic assessments, allowing people from all over the world to access knowledge and learn as if they were in regular classrooms. Although meditation-based interventions are not simply a form of knowledge, the potential of online courses of meditation-based interventions should be explored for clinical research and applications. Commercial products aiming to create personalized experiences for meditation practitioners are on the rise, ranging from mobile applications that record and prompt daily meditation practices to simple feedback devices that seek to facilitate individuals in achieving a meditative state. If one were to search the phrase *personalized meditation* in Google Scholar one would notice that there are quite a few patent registrations for personalizing meditation through simple monitoring devices and mobile applications. Therefore as researchers we need to start exploring these opportunities that may potentially benefit our research and applications (see Chapter 10: Critical questions and future directions in meditation).

What can be done to ensure the quality and effectiveness of online courses? This issue was brought up previously in our discussion of web-based meditation programs (see Chapter 8). The first necessary step is obviously to standardize these online meditation-based interventions based on the offline standardized programs. The same curriculum and structure should be followed and used for online interventions. Referencing the range of online courses offered by notable universities, video recordings of lectures are always the major form of communication between students and instructors. Although real-time feedback and questions are often missing in these online courses (i.e., lectures are prerecorded), it does not mean that real-time interaction is impossible. For those of us who engage in video conferencing, comments and questions can often be submitted to the host of the conference, allowing for real-time communication and feedback. During a typical face-to-face group-based intervention course, participants usually sit in a circle with the instructor. The instructor would guide the participants during practices and engage them in discussion. All of these necessary offline course components can be equally realized via online platforms. The success of online college courses suggests that it is feasible to adapt group-based intervention into online courses as long as we also make these courses as closely matched as possible to face-to-face intervention. Importantly, once we enable the feature of real-time communication and allow instructors to observe each participant through video cameras, we would further reduce the differences between online and offline group-based intervention programs. Nevertheless, unforeseen technical issues and other practical problems would likely arise while we transform these interventions online. Therefore in order to realize the promises of personalized medicine for meditation-based interventions, there is an urgent need for more rigorous and large-scale clinical trial studies examining the feasibility and effectiveness of online meditation-based interventions in improving different psychological and cognitive outcomes.

Personalized meditation and provider attention

The uniqueness of personalized experience often involves receiving undivided or dedicated attention from the experience providers. For individual-based psychotherapy the experience is often a "personalized" one in that patients have the full attention from their therapists and can communicate their concerns and have those addressed in a timely fashion. However, meditation-based interventions are rarely taught between one instructor and one student, instead, a group of individuals are taught simultaneously by an instructor. The attention resources of the instructors are often divided into group-based teaching scenarios and cannot be divided equally to each participant in the group. If we plan to personalize meditation practices and tailor them to individual needs and preferences, then it is important to think about how to also personalize the attention an individual receives during the intervention, or how can instructor's attention be better devoted to individual participants or patients. The most straightforward solution is to transform personalized meditation into an individual-based experience, just like psychotherapy. There are no obvious reasons against providing individual-based meditation intervention. In many ancient traditions, the teaching of meditation often occurs between a student and a teacher. However, there is a potential issue of increasing healthcare cost for this kind of individual-based approach to meditation, which could be a barrier for transforming group-based intervention into a personalized one. Another solution to this issue of instructor's attention would be to have additional individual-based sessions on top of the standard group-based intervention. For instance, instructors could at least have one face-to-face session with each individual participant to have more in-depth and personalized discussion regarding problems, questions, and concerns that have arisen during individual's practices. Currently most standardized group-based meditation interventions are missing this personalized interaction and discussion component with the instructor, which could potentially be problematic for addressing adverse experiences and effects in a timely manner. For example, if individuals encounter adverse experiences and effects during their practice of meditation, they may not be willing to share their concerns and experiences with other people in the group. Yet, without a dedicated opportunity to discuss their problems with the instructor, individuals may have to deal with the adverse experiences themselves, which could potentially exacerbate their problems, or they may drop out of the intervention altogether. Therefore personalizing group-based interventions, not just meditation-based intervention, can potentially be beneficial for intervention effectiveness and the patient's satisfaction and well-being.

Personalized meditation and provider awareness

When healthcare professionals think within the framework of personalized medicine, they are more appreciative of its potential in improving patient outcomes. However, they may not be the one to actually provide these personalized treatment approaches to individual patients and may only be involved in the early stages of diagnosis and treatment planning. As a result the providers of these personalized treatment approaches would need to be trained under the framework and principles of personalized medicine and be able to adapt their treatment strategies and habits to better fit the mission of personalized medicine. For meditation-based interventions, this change of mindset and awareness is especially important for intervention providers (typically the instructors of meditation practices) since these instructors are used to teaching the practices in group-based settings and may not tend to engage heavily with individual participants on a one-to-one basis. Based on our discussion of potential ways of personalizing meditation-based interventions, instructors may still teach meditation in a group-based setting, but the focus should now be directed more toward each individual participant instead of the group as a whole. Furthermore, when instructors interact with individual participants they should be trained to gear their instructions and communicative styles to individual participants' needs and characteristics. In some sense, instructors involved in personalized interventions would need to adopt a therapist's mindset to facilitate their communication with individual participants. In order for this transformation to happen, collaborations among clinical psychologists, therapists, meditation instructors, and researchers are necessary to develop guidelines for effective implementation of personalized intervention approaches to each individual.

Personalized medicine and administrative barriers

The potential and promises of personalized medicine cannot be further emphasized, but realizing and implementing these personalized treatment approaches can encounter barriers that are out of the control of researchers and healthcare professionals. For any kind of new drugs, therapies, technologies, or interventions, there are many administrative and governmental hurdles to jump through before these approaches can be utilized in medical and clinical settings. This is not to say that we should implement treatment approaches without examination and approvals, but this approval process can often take longer than it should. Prioritizing personalized medicine approaches for administrative approval would greatly incentivize researchers and commercial companies into developing new diagnostic and treatment tools to further improve the efficacy of existing healthcare services (Aspinall & Hamermesh, 2007). In addition to getting approvals for the application of personalized interventions, the cost of these personalized approaches should also be taken into consideration. If we develop interventions that are not affordable or are not covered by health insurances, then patients are less likely to benefit from these approaches because of the high cost. The financial barrier should also be addressed in advance by communicating with insurance companies with regard to the percentage of reimbursement if patients plan to choose personalized interventions. This concern of cost is not trivial, as nearly 80% of medical bills in the United States are covered by insurances, Medicare, or Medicaid (Aspinall & Hamermesh, 2007). If we cannot find affordable ways to pay for personalized treatment approaches including personalized meditation, then implementing them would also be meaningless since the benefits would most likely only reach a small percentage of the population rather than the majority.

Summary and implications for translational research

In this chapter we discussed the proposition of personalizing meditation-based interventions to better meet the needs and characteristics of individuals. The idea of personalizing meditation is heavily inspired by personalized medicine, which has gained traction from governmental agencies and scientific community over the past decade. We described how to develop personalized treatment approaches by first introducing the importance of making predictions about individual responsiveness to a host of interventions and treatment. Based on existing evidence and theories, personality traits, genomics, physiological indexes such as brain waves and heart rate variability, and preexisting individual differences in psychological well-being and cognitive function can all be useful predictors of intervention responsiveness. This prediction would allow us to tailor intervention approaches to the needs of individuals. We then proceeded to discuss how to implement personalized intervention approaches in clinical settings, highlighting the challenges and barriers in realizing the promises of personalized interventions. Notably, we discussed the importance for healthcare providers to think within the framework of personalized medicine in order for such approach to be effective. Furthermore, we explained that implementing personalized interventions including personalized meditation would require more rigorous feasibility studies and clinical trials to evaluate the effectiveness of these approaches. From a translational research and application perspective, personalized meditation indeed holds promise for improving the quality and effectiveness of existing meditation-based interventions in that individual preferences and needs are better attended than conventional approaches. However, both prediction and implementation need to be informed by scientific research, which is currently in the infancy stage since not much has been explored with respect to individual differences in response to meditation-based interventions. For future investigations it is critical to advance our understanding of individual-specific characteristics and factors that may play a role in affecting intervention responsiveness. Moreover, exploring ways to effectively and efficiently develop and adapt existing intervention approaches for personalization would be highly important for clinical applications.

References

Anderson, T., & Farb, N. (2018). Personalizing practice using preferences for meditation anchor modality. *Frontiers in Psychology, 9*, 2521.
Aspinall, M. G., & Hamermesh, R. G. (2007). Realizing the promise of personalized medicine. *Harvard Business Review, 85*(10), 108.

Beevers, C. G., & McGeary, J. E. (2012). Therapygenetics: Moving towards personalized psychotherapy treatment. *Trends in Cognitive Sciences*, *16*(1), 11–12.

Chapman, B. P., Roberts, B., & Duberstein, P. (2011). Personality and longevity: Knowns, unknowns, and implications for public health and personalized medicine. *Journal of Aging Research, 2011*.

Conway, A. R., Kane, M. J., & Engle, R. W. (2003). Working memory capacity and its relation to general intelligence. *Trends in Cognitive Sciences*, *7*(12), 547–552.

Davidson, R., & Kabat-Zinn, J. (Eds.), (2004). Letters to the editor. *Psychosomatic Medicine*, *66*, 149–152.

Edwards, E. (2012). The role of complementary, alternative, and integrative medicine in personalized health care. *Neuropsychopharmacology*, *37*(1), 293.

Eley, T. C., Hudson, J. L., Creswell, C., Tropeano, M., Lester, K. J., Cooper, P., . . . Uher, R. (2012). Therapygenetics: The 5HTTLPR and response to psychological therapy. *Molecular Psychiatry*, *17*(3), 236.

Fingelkurts, A. A., Fingelkurts, A. A., & Kallio-Tamminen, T. (2015). EEG-guided meditation: A personalized approach. *Journal of Physiology-Paris*, *109*(4-6), 180–190.

Hamburg, M. A., & Collins, F. S. (2010). The path to personalized medicine. *New England Journal of Medicine*, *363*(4), 301–304.

Lester, K. J., Hudson, J. L., Tropeano, M., Creswell, C., Collier, D. A., Farmer, A., . . . Eley, T. C. (2012). Neurotrophic gene polymorphisms and response to psychological therapy. *Translational Psychiatry*, *2*(5), e108.

Lester, K. J., Roberts, S., Keers, R., Coleman, J. R., Breen, G., Wong, C. C., & Eley, T. C. (2016). Non-replication of the association between 5HTTLPR and response to psychological therapy for child anxiety disorders. *Brithish Journal of Psychiatry*, *208*(2), 182–188.

Smith, J. (2004). Alterations in brain and immune function produced by mindfulness meditation: three caveats. *Psychosomatic Medicine*, *66*, 148–149.

Tang, Y. Y., Hölzel, B. K., & Posner, M. I. (2015). The neuroscience of mindfulness meditation. *Nature Reviews Neuroscience*, *16*(4), 213.

Walsh, J. J., Balint, M. G., Smolira, D. R., SJ, Fredericksen, L. K., & Madsen, S. (2009). Predicting individual differences in mindfulness: The role of trait anxiety, attachment anxiety and attentional control. *Personality and Individual differences*, *46*(2), 94–99.

Chapter 10

Critical questions and future directions in meditation

Meditation has gained traction among researchers and the general public for nearly three decades. Although contemplative practices themselves have been around for thousands of years, modern adaptation of meditation practices has provided an interesting twist on these ancient techniques and practices. We have been relatively successful in understanding what kind of behavioral effects meditation has on aspects of psychological health, well-being, and cognitive function. We have also been up to speed with the latest technological and methodological advances from other fields of study and have actively exploited these methods in our research to further reveal the effects and mechanisms of meditation at physiological, biological, and neural levels. Our efforts have, without a doubt, greatly impacted our scientific knowledge and applications of meditation in promoting health and well-being. However, misconceptions and overinterpretation of meditation effects as well as methodological issues in meditation research have cast a heavy shadow over this promising contemplative practice. Just like any other field of research, the process of obtaining comprehensive and accurate knowledge is always complex and ongoing. For researchers and public consumers of scientific findings, we would like to highlight and discuss critical questions that the field of meditation and contemplative practices would need to address in future investigations. Furthermore, we would like to discuss promising research directions in light of the increasing technological advances in the digital age that have great potential in enhancing the experience and effects of meditation.

Instructor (trainer) effects on meditation outcomes

The first critical question is how important instructors of meditation are with respect to intervention effects and outcomes. Generally we focus on whether or not an instructor is certified to teach meditation or how long their teaching and practice experiences are to make judgments about their qualification. These basic criteria are necessary and common in education, yet we would likely recall a particular instructor, coach, or teacher who we think are better

The Neuroscience of Meditation. DOI: https://doi.org/10.1016/B978-0-12-818266-6.00011-3

in teaching the materials than anyone else. Why? How are they different? If we reflect on our reasons for selecting a particular individual as the best instructor, the reasons that come into mind are most likely to be more than just "they are knowledgeable" or "they communicate things clearly." For example, some of us may like the personality of our favorite instructor, while some of us may favor his or her unique teaching style. Whichever it is (including other possibilities we did not mention), we would unequivocally agree that our favorite instructor had a positive impact on our learning and, most likely, the outcomes of learning. In education, our emphasis is always more on certification, which is critical and fundamental factors enabling high-teaching quality and outcomes. However, there are also individual characteristics specific to the instructor that may interact with the teaching process even when instructors are trained to follow a standardized manual or curriculum. Taking the personality of the instructor as an example, an extraverted instructor is likely to interact more with their trainees and may also foster a relatively lively learning environment that promotes motivation, enthusiasm and engagementfor learning among the trainees. These motivational factors may subsequently improve the outcomes of learning.

Let's use an analogy to help better understand the subtle instructor effects on learning outcomes. If we like margherita pizzas, we often know where to get the best margherita pizza in town. We know exactly which restaurant has the best margherita pizza we want, and it is the best compared to any other restaurants we have tried that offer the same kind of pizza. If all the restaurants use the same ingredients such as cheese, tomato sauce, and basil and the chefs follow the same recipe and procedures to prepare our pizzas, why is our taste of the same kind of pizza different? While some of us may say that there have to be some subtle differences in steps that cause these differences in taste—just think about what is it like when we follow a particular recipe by a celebrity chef, we often do not end up with the same delicious dish. Sometimes the differences may be due to errors, while at other times it may be in subtle details not described in the recipe, but are nevertheless crucial to the taste of the dish. All of these differences are natural as it is not possible for us to do things as exactly the same as everyone else. Therefore in the context of instructors or trainers of meditation, their individual characteristics and ways of teaching could potentially be different from one another and could play an important role in affecting the learning and outcomes of meditation (Tang, 2017).

Similarly, within psychotherapy, experienced therapists are likely to have better patient outcomes than therapists who have just entered the practice with certification. For meditation intervention or training, an experienced instructor or trainer who has repeatedly achieved a meditative state would know how to best direct and interact with trainees to help them achieve the desired meditative state. Experienced instructors are also likely to have similar experiences and problems when they first started to practice and have accumulated experiences on how to effectively handle certain situations that may arise during the stages of learning. An experienced instructor can also give immediate feedback (like a mirror function) during and after the intervention sessions to facilitate the trainees to practice effectively and correctly. It should be noted that when we talk about experiences, we are not referring to the amount of training an instructor needs to have before he or she could teach meditation to other individuals. Instead, we are referring to experiences that he or she accumulated during the process of teaching. These experiences could potentially play a critical role in how instructors go about their teaching even under the guidelines and framework of a standardized curriculum. However, individual instructor effects on learning and intervention outcomes are often not examined in meditation studies and virtually no data are collected from the instructors themselves. Therefore we are virtually in the dark with regard to their roles in affecting the practice and outcomes of meditation. Nevertheless, there are reasons to expect that an experienced instructor could facilitate and support the trainees more efficiently than an instructor who just begins to teach the intervention, and that instructors could well be an active component of meditation interventions.

For instance, if we look into organizational behavior and leadership research, a leader has a direct impact on group dynamics, subsequently influencing group members' motivation and outcomes (Fox, Rejeski, & Gauvin, 2000). In particular, the personality of the leader, a determinant in leadership styles, also has critical influence on the team dynamics, which, in turn, affect organizational performance (Peterson, Smith, Martorana, & Owens, 2003). In medicine and psychotherapy, the physician–patient or therapist–client relationship such as trust and intimacy play key roles in treatment acceptance, compliance, and outcomes (Butler & Strupp, 1986; Frank, 1959; Truax & Carkhuff, 2007). Likewise, the interaction between the instructors and trainees can also form group dynamics that are subject to evolve and behave according to the principles of organizational behavior, especially when a group of trainees practice together under the guidance of the instructors. Therefore a promising and potentially influential research area for meditation research would be to consider the degree to which meditation learning and outcomes can be influenced by the instructor's individual characteristics and personality, as well as the overall group dynamics. Joining efforts with researchers from industrial and organization behavior would be essential in uncovering this overlooked aspect of group-based intervention or training programs, as currently there are very limited studies that addressed this topic in the meditation field.

Meditation practice is a learning process and learning capacity can influence learning processes, thereby affecting practice/treatment processes and outcomes. So far there is no study that addresses this question, but a recent review article has explored the potential role of learning capacity in cognitive behavior therapy (CBT) for depression and may provide insight for

194 The Neuroscience of Meditation

meditation learning. As we know, CBT is one of the most effective and best-tested treatments for depression. However, only about 50% of the patients respond to CBT and at least 25% of the patients who respond to treatment relapse within a year (Lemmens et al., 2019). Since CBT is a skills-based approach, the role of learning and memory in CBT is important for individuals to grasp the taught skills, but the exact role of learning in relation to CBT remains unclear (Bruijniks, DeRubeis, Hollon, & Huibers, 2019). Nevertheless, these findings of CBT can inform us about meditation learning. First, the role of learning in CBT for depression suggests learning processes are involved in the development and maintenance of depression. Regarding meditation, prior learning experiences and strategies are important for later learning and practice, which may help instructors determine the appropriate teaching strategies to better fit the learning capacity of participants. Second, learning capacity moderates the relation between CBT procedures and change in CBT treatment processes and explains why therapeutic procedures lead to process change and long-term success in some, but not all, patients. The same is likely to be true for meditation-based intervention, suggesting that instructors should consider all these factors, identify core procedures that lead to successful learning, and tailor the experience for each individual to achieve the intervention goals more effectively.

Different meditation techniques matter

When we think about standardized meditation programs, we would immediately realize that a range of different meditation techniques are included in the curriculum. There are also obvious differences in these techniques, not to mention the behavioral effects, in terms of brain functional activation while each technique is being practiced by the individuals (Fox et al., 2016). Research has begun to look at the psychological and cognitive effects of individual meditation techniques and has demonstrated distinct brain regions involved in each of these techniques (Fox et al., 2016; Lippelt, Hommel, & Colzato, 2014). Taking focused attention and loving kindness as examples, we know that the former involves focusing and maintaining attention on a specific object throughout the practice, while the latter involves fostering a sense of love and kindness toward oneself and others. For focused attention, the active engagement of emotion is missing, whereas for loving kindness, a heavy engagement of attention is missing. These two techniques have been shown to engage different brain regions, which indicates that they are also likely to induce different intervention outcomes with respect to attention and emotion. As Lippelt et al. (2014) have pointed out, we currently have a poor understanding of the differences in outcomes as a function of each meditation technique. Given that most standardized meditation intervention encompasses a range of techniques, it is difficult to tease apart the role of each technique in contributing to the observed behavioral and brain outcomes.

Relatedly, from translational and clinical perspectives, these topics are relevant for helping individuals achieving the most desirable outcomes. For instance, if we have a patient with attention-related deficits, would it make sense to teach this patient both loving kindness and focused attention techniques, or should we just teach this patient focused attention and emphasize this specific practice as the most relevant to their behavioral problems? The answer is likely to be that focused attention should be the emphasis, yet we are lacking empirical support for this logical assumption. It is also worth considering the fact that individuals may have preferences for meditation techniques, such that they may prefer to practice one technique over other alternatives. In our daily life, we often have specific preferences for food, taste, and drinks. Our unique preferences and individual characteristics guide our behavior at every moment. It is not unreasonable to expect that when we have preferences for Coca Cola or Pepsi, we would also have preferences among meditation techniques.

When we examine the psychotherapy literature, we can find empirical support for individual preferences and, more importantly, it is critical for therapists to be attentive of these individual preferences, as poorer outcomes of intervention are likely to occur if these preferences are not adequately addressed (Williams et al., 2016). One preliminary report has indicated that there are preferences for specific meditation techniques and not all techniques are perceived as equally desirable by individuals, suggesting that people may not find every meditation technique helpful and may only choose to practice their preferred techniques (Burke, 2012). Relatedly, preferences for specific modality of mindfulness practice anchors (e.g., using breath, imagery, or auditory-phrase as a focus of attention) are shown among novices who practice mindfulness and such preferences can undergo change over the course of mindfulness meditation intervention (Anderson & Farb, 2018). These preferences would likely influence the ability of individuals in achieving optimal meditation states and outcomes in that practicing meditation through their most preferred techniques and modality would likely increase their chance of entering the meditative state. Hence different meditation techniques are likely to have differential impacts on certain outcomes, both through their distinct focuses and strategies to obtain meditation states and individuals preferences for specific techniques.

Effort and stage of practice matter

Another important, yet subtle, aspect of meditation practices is the amount of effort used in achieving the meditative states. Even within the same meditation technique, there are differences in teaching emphasis and guidance with respect to the amount of effort an individual should use to maintain focus on the chosen target. For example, in practice, the amount of effort has been conceptualized as effortful practice and effortless practice (Tang, 2017). Effortful practice involves a considerable amount of effort devoted to maintaining the attention focus throughout the meditation practice, and vice versa for effortless practice. The reason for exerting a lot of effort is often due to the high tendency of mind-wandering, which is typical in the early stage of practice. Novices who just begin the practice are likely to engage in effortful meditation, whereas those who are more experienced, or at an advanced stage of practice, tend to use minimal amount of effort to stay focus as their tendency to wander away from the present moment has greatly decreased. This difference between effortful and effortless practices could be one of the key features distinguishing expert meditators from novice meditators. Hence the amount of effort used in meditation practice is worth being operationalized into an objective construct, which can be used in our empirical investigations of meditation.

Another closely related topic is the formal and informal practices of meditation, which refers to practices that either happen within intervention settings and are often guided by instructors (formal practice) or happen outside of intervention settings in which individuals voluntarily engage in self-practice. This is also dependent on the stage of meditation. In the early stage, regular and formal practice (e.g., fixed time, location, environment) is very helpful for individuals to build new habits and form new synaptic connections within the brain to reconsolidate the learning. In the advanced stage, daily informal practices may be more desirable than formal practices since the meditators have formed their own habits and preferred ways of practice and are less likely to need step-by-step guidance as they had before. However, it is also likely that meditations in advanced stages would be fine with both formal and informal practices. Different stages of meditation and expertise may also be related to the use of different techniques in achieving the meditative states. For example, a novice meditator may focus on one meditation technique and strive to become skillful in this one technique, and then move toward practicing other techniques. However, it is possible that a novice meditator may feel uncomfortable and even upset with this one technique and decides to quit (e.g., a novice may think that "I'm not suitable for meditation as I could not maintain my attention on my object of focus"). One option may be to let them try out other techniques that may be more suitable for them. This option is related to our discussion of individual preferences, indicating that determining preferences or finding ways to predict individual preferences could be more efficient than going through a series of trials and errors before deciding on the most suitable technique. Another option is to adjust the teaching and practice guidance to see whether these changes alter the individual's approach to the practice. For instance, an expert instructor could guide the trainee by asking, "No matter whether you achieve your goal or not, if you just focus on your breath rather than your feeling in which you are judging, what do you feel? Do you still want to quit the practice?" If this still does not work and the individuals are open to other potential new techniques, then selecting another appropriate meditation technique becomes crucial. This stage of frustration can even be found among advanced meditators who could also be stuck during the meditation process and feel bored and unhappy in what he or she has been doing for a long period of time. An experienced instructor would be able to notice this struggle and understand the underlying and hidden reason for this frustration as she or he may have undergone a similar stage. Subsequently, an experienced instructor can figure out solutions based on his or her own experiences and motivate the advanced-level trainees to continue their practices.

These different stages of meditation practice are also largely unexplored, since we often assess individuals at baseline and at intervention completion. Intermediate assessments when the intervention is still ongoing are rare, but may nevertheless be informative for tapping into the dynamic process of meditation-related changes, as well as potential changes in the stage of meditation, such that an individual may progress from being a complete novice to a more advanced stage meditator. This idea of meditation stages and the change in stages at the individual level is related to our emphasis on interindividual differences with respect to changes from one time point to another time point. More importantly, treating the individual meditator as a unit of investigation is likely to provide more information regarding how each individual varies as opposed to how a group of meditators vary from baseline to postintervention. Overall, effort and stage both can be interesting investigative targets of meditation research, but currently we do not have a clear understanding of how these two factors may interact with meditation practices and outcomes.

Meditation in the digital age

For better or worse, the use of technology is ubiquitous and necessary for people living in the digital age. Although the use of mobile phones, computers, and artificial intelligence have greatly improved our quality of life, they also overwhelm us with information. While everything has two sides, one positive, and one negative, it is important to exploit these technological advances to our advantage with regard to their potential in promoting the effects of meditation. We have discussed the potential benefits of use of real-time neurofeedback to monitor and improve the quality of meditation practices at the individual level. However, having Electroencephalography (EEG) neurofeedback (NF) equipment at home is not always realistic and feasible as this kind of equipment is often designed for scientific use and is usually prohibitively expensive. However, there have been some portable devices on the market that claim to provide some level of NF for meditation practitioners. Although it is difficult to know whether such equipment is effective and accurate in providing real-time NF, the idea is a very attractive one. If we utilize NF in laboratory setting in combination with meditation intervention and find this combination to be effective, we would need to come up with ways to have this technology benefit the general public, otherwise it would be meaningless if we cannot translate this NF technology into action. Therefore developing more feasible ways of NF for everyday use is also a promising research direction that requires collaboration across different disciplines.

There are more ways to utilize technology to facilitate our meditation practices in the digital age. Going back to our brief discussion of mobile apps on smartphones, which include audio guidance for different meditation practices and techniques, we can build and develop our intervention programs through these virtual platforms, allowing wider dissemination of intervention while also reducing the human power and resources that are often nontrivial in face-to-face intervention programs. We can reach a wider audience through online platforms (Davidson & Dahl, 2018). We can also collect behavioral data such as questionnaires and even cognitive tasks through smartphones and tablets, which are evidently more efficient than having individuals coming into a laboratory to participate in different assessments. However, there are definitely challenges to provide interventions via online platforms. The standardization of online meditation interventions has yet to be made or thoroughly investigated. Moreover, interpersonal interaction is unlikely to be more direct compared to face-to-face intervention. Relatedly, if group dynamics are an important part of meditation intervention, then it is also going to be different within online platforms. Although it is always possible to have group-based meeting and interaction online, how exactly this format of interaction would affect intervention outcomes are still unknown. Therefore feasibility studies of online meditation interventions would be critical for any further development of online programs. Moreover, scientific investigations of the effectiveness of online programs on common meditation outcomes should not be carried out unless there is a feasible online meditation program established in place. Nonetheless, moving toward a broader and wider dissemination of meditation-based intervention would likely have positive impacts on society as a whole.

Combining meditation with other intervention approaches

From a translational and clinical perspective, we always seek to improve the efficacy and effectiveness of existing interventions. In scientific research and clinical applications we primarily focus on understanding the effects of one intervention on the brain and behavior in different populations. However, one interesting prospect of research on interventions is that we can also potentially combine or integrate interventions together to improve the overall effectiveness and efficacy in improving our targeted outcomes. For instance, considering the possibility of combing meditation intervention and physical exercises/training. We know that meditation intervention mostly involves mental practices and seldom has physical activity. Even when physical activities are engaged (like in mindfulness-based stress reduction) they are mostly moderate movements such as yoga poses and stretching. However, it is important to realize that even though physical activities are not heavily emphasized in meditation-based interventions, they are still being included in some of the standardized programs, suggesting that incorporating mental practices and physical activities into an intervention program is most likely to be feasible. Based on research of physical exercises and training, its benefits on physical and psychological health have repeatedly been shown and many of the observed benefits often overlap with those found in meditation research (Penedo & Dahn, 2005). For most people, physical exercises are common in their everyday life and well-perceived by the general population. Given that one intervention may be more active and another more static, their complementary nature may be well-suited for integrating and combining into a new intervention that may potentially induce greater benefits on psychological and physical health. There have not been a lot of studies on combining meditation-based interventions with other forms of training or intervention. However, this approach could be promising for enhancing the overall effectiveness of interventions and is worth investigating in future studies.

In addition to this example, we will take EEG NF as an example that illustrates the utility of integrating different interventions. NF is a noninvasive tool that targets certain brain networks for neuromodulation by including real-time measures of brain activity, thereby enabling deliberate self-control over specified EEG frequency bandwidths through such feedback (Gruzelier, 2014). Research studies have suggested that NF may contribute to learning more effectively through self-control of mental processes, which may include processes that are commonly engaged in meditation, such as attention, emotion, and self-awareness. For novices who often have difficulty grasping how to meditate effectively, NF can provide an individualized real-time feedback and instruction by linking objective brain states to subjective meditation experience, which can help novices to learn and adjust their attention and effort more effectively. Imagining a scenario where an individual can practice meditation along with real-time NF, whenever his or her attention starts to wander, NF would prompt the individual and help them adjust back to a proper meditation state. This would undoubtedly be beneficial for novices who tend to experience mind-wandering during meditation practices by reducing the amount of time they spend on mind-wandering through prompt feedback. Therefore NF may provide an effective way to augment meditation effects and improve intervention efficacy and overall engagement. However, how to design an effective NF protocol to combine with meditation is crucial (Marzbani, Marateb, & Mansourian, 2016). Based on evidence from our studies and that of other groups, although the results of NF using different brain frequency waves seem mixed, most people seem to benefit from NF using theta and/or alpha bandwidths, consistent with the EEG findings in meditation research. Therefore for any empirical studies that seek to use NF in combination with meditation, evidences suggest midline frontal theta (FMθ) NF as the main target (Brandmeyer & Delorme, 2013; Enriquez-Geppert et al., 2014). Moreover, many studies have shown that FMθ in the anterior cingulate cortex (ACC)/medial prefrontal cortex is mostly involved in high-level self-control and are also regions in self-control networks commonly activated in meditation. These lines of evidence further suggest the possibility of augmenting meditation effects via NF, subsequently improving the engagement and quality of practice than meditation alone. In a pilot study, our group used FMθ NF to maximize the efficiency of integrative body−mind training (IBMT), a form of mindfulness meditation, and revealed some promising trend of improvement, while also demonstrating the feasibility of combining NF with meditation. Further research should explore the effects of combining NF and meditation intervention on brain and behavioral outcomes, develop optimal meditation + NF protocols, and investigate the underlying brain mechanisms following meditation + NF in order to maximize the effectiveness of meditation intervention in achieving different goals such as stress reduction, enhancing attention, and well-being.

Differences among meditation, psychedelic, and other states

The phenomenological and neurophysiological research of meditation and psychedelic drugs has been growing over recent years. Although both approaches induce the altered states of consciousness (ASC) that are mainly mediated by agonism of serotonin receptors, the similarities and differences between these states and their underlying mechanisms remain unclear. For example, studies often focused on one ASC phenomenon——ego dissolution—because meditation practice often produces the sense of self-dissolving, whereas psychedelics often produce disruptions of self-consciousness or drug-induced ego dissolution. We will discuss some available evidence regarding similarities and differences between meditation practice and psychedelic drug-induced states using phenomenological and neurophysiological data, with a particular emphasis on alterations of self-experience. We also compare the states of meditation and hypnosis, as well as their different mechanisms.

A pilot study examined whether the psychedelic drug psilocybin could assist mindfulness meditation and modulate self-consciousness and brain default mode network (DMN) connectivity, as DMN is involved in the sense of self. The logic was that both psychedelics and meditation produce profound effects on self-consciousness, perception, emotion, and cognition and their combination may have synergistic effects on brain networks. During a 5-day mindfulness retreat, participants received a single administration of psilocybin (psilocybin-assisted mindfulness session). The resting-state functional connectivity showed psilocybin-related and mental state-dependent alterations in self-referential processing areas of the DMN, similar to the DMN changes following meditation alone. Decoupling of DMN in medial prefrontal and posterior cingulate cortexes was associated with the subjective report of ego dissolution during the combined session. The extent of ego dissolution and brain connectivity predicted changes using a self-report questionnaire which measured changes in attitude in the psychosocial domain four months later. One question is that the study did not use commonly accepted psychosocial measures to evaluate the before−after changes, which are difficult to compare with published results in the field. Therefore it remains unclear whether combined psilocybin and meditation could indeed promote positive behavioral changes. Given that the sample size was small and all participants were long-term meditators in a retreat environment, the retreat setting with intense meditation practice may serve as a supporting environment that contributed to the self-report and DMN changes. Further research is warranted to replicate these preliminary results (Smigielski, Scheidegger, Kometer, & Vollenweider, 2019).

Converging evidence suggests that high doses of hallucinogenic drugs can produce significant alterations of self-experience——the dissolution of the sense of self and the loss of boundaries between self and world. This phenomenon is known as drug-induced ego dissolution or self-dissolving. Interestingly, meditation practice also leads to a similar self-experience of oneness, self-dissolving, or nonboundaries between self and world. However, the meditator is still aware of and has control of the changes of these mental states, which is significantly different from the states controlled by drugs (Tang & Tang, 2014). Self-report questionnaires suggest that at least three types of drugs can induce ego dissolution: classical psychedelics, dissociative anesthetics and agonists of the kappa opioid receptor. Although these drugs act on different neurotransmitter receptors, they all produce subjective effects comparable to the symptoms of acute psychosis including ego dissolution. Neuroimaging studies have suggested that brain processes of drug-induced ego dissolution may indirectly reveal the neural correlates of the self. However, neural correlates of ego dissolution might reveal the necessary neurophysiological conditions for the maintenance of the sense of self, but it could reveal its minimally sufficient conditions (Millière, 2017). A recent review summarized the consequences of psychedelics on ASC such as profound alterations in sensory perception, mood, thought (e.g., the perception of reality), and the sense of self, as well as the progress in the search for invariant and common features of psychedelic states. Although the possible neural mechanisms of acute effects of psychedelics on sensory and time perception as well as emotion and cognition have been shown, the influence of nonpharmacological factors on the acute psychedelic experience such as demographics, genetics, personality, and mood are also important. In the study, since psychedelic participants showed different stages over time and levels of changes along a perception-hallucination continuum of increasing arousal and ego dissolution, the authors proposed that psychedelic experiences have a dynamic nature. Future research is needed to better understand different short-term or long-term psychedelic states of consciousness and their underlying brain mechanisms (Preller & Vollenweider, 2018).

While meditation and psychedelics may change or disrupt self-consciousness and underlying neural processes, it seems that neither meditation nor psychedelic states can be regarded as a simple state of consciousness. Moreover, there are important phenomenological differences even between conscious states like experiences of self-loss or self-dissolving. As a result, self-consciousness is best construed as a multidimensional construct and that self-loss can take several forms. For instance, various aspects of self-consciousness, including narrative aspects linked to autobiographical memory, self-related thoughts, mind-wandering, and embodied experiences rooted in multisensory processes may be differentially affected by psychedelics and meditation practices. Most importantly, what are the long-term outcomes of experiences of self-loss induced by meditation and psychedelics on personality traits and social behavior? We should also carefully consider the similarities and differences between states and traits, (e.g., temporary states of self-loss, self-dissolving, or ego dissolution is very different from the behavioral or social trait of selflessness), although there is preliminary evidence of correlations between short-term experiences of self-dissolving and long-term trait alterations (Millière, 2017; Millière, Carhart-Harris, Roseman, Trautwein, & Berkovich-Ohana, 2018).

Relatedly, some believe that other mental states such as the hypnosis state is similar to the meditative state and even think meditation is a form of self-hypnosis. We here consider hypnosis and meditation to compare their effects on executive attention——a form of self-control capacity—and delineate the similarities and differences between these two states in modifying the ability of conflict resolution (Tang & Posner, 2015). The Stroop and flanker tasks are among the most common ways to study the resolution of conflict. Stroop interference refers to the response-time difference between congruent and incongruent trials when participants are required to respond with the ink color of a color word (i.e., the word "blue" in the congruent color "blue" or the word "blue" in the incongruent color "red"). Although there are many ways to administer the Stroop task, they are all based on a comparison between congruent and incongruent trials. Longer reaction time (RTs) and higher error rates on incongruent trials are a reliable feature of the task, at least for people skilled in reading. A second way to examine brain responses to conflict resolution is the flanker effect in which response time to a target surrounded by congruent flankers (same directions) is compared to response times when the flankers are incongruent (different directions). Could both hypnotic suggestion and meditation training influence behavioral performance in conflict tasks by reducing the time to resolve conflict?

In general, hypnosis is not a training method, rather it is a highly focused absorbed attentional state that minimizes competing thoughts and sensation. This state can result in stronger control of behavior from an outside force (e.g., a hypnotist). The use of posthypnotic suggestion allows the hypnotist to alter the behavior of a person even when they are not in the hypnotic state (Oakley & Halligan, 2013). Studies indicated that the Stroop effect can be greatly reduced or abolished by inducing a hypnotic state in highly suggestible people and giving the suggestion that the word is a meaningless string. For example, one study found that the Stroop effect in reaction time declined from 91 to 9 ms when highly hypnotizable participants were given the suggestion. Without the suggestion there was no significant change. However, for less hypnotizable participants the suggestion reduces Stroop interference only slightly, suggesting hypnosis only works for highly hypnotizable participants (Raz, Fan, & Posner, 2005). Here we take IBMT, a form of mindfulness meditation, as an example to reveal the Stroop effect induced by meditation. In our randomized controlled trials, we first found that five sessions of IBMT (30 min per session) can improve performance on conflict resolution using the Flanker task. These results led us to expect a similar influence on the Stroop task (Tang et al., 2007). To determine the time course of conflict resolution and its brain mechanisms before and after training, we randomly assigned participants to an IBMT group or to a RT group. Prior to and following 10 sessions of training (30 min per session, 5 h in total), all subjects performed the Stroop task while their brain activity was recorded using event related potentials (ERPs). We hypothesized that after training, IBMT would reduce the Stroop interference effect in reaction time more so than the RT. Consistent with our hypothesis, our results showed a highly significant improvement in Stroop interference in the IBMT group (30 ms) versus a nonsignificant improvement (6 ms) in the relaxation control group; a similar result was also obtained for behavioral accuracy. However, even after meditation training, participants still showed less time for resolving conflict, suggesting the lasting effects of meditation. This stands in contrast with the nonsignificant differences in time for conflict resolution in highly hypnotizable participants after posthypnotic suggestion (Fan, Tang, Tang, & Posner, 2014, 2015).

Do hypnosis and meditation involve the same brain mechanisms of conflict resolution? Combining functional magnetic resonance imaging (fMRI) and ERP could give us a picture of the time course and localization of changes in the brain following hypnotic suggestion or meditation training. Results showed that posthypnotic suggestion greatly reduced the fMRI activity and ERP from posterior brain areas including the ERP components P1 and N1 visual responses (Raz et al., 2005). It also virtually eliminated the usual increased negativity for incongruent trials (N2 component) over frontal midline electrodes that has been thought as conflict-related activity of the ACC and greatly increased the latency of this difference in the subsequent positivity (P3 component) reflective of cognitive effort. These results suggested that during hypnotic suggestion the word is treated as a meaningless string and the visually presented word produced less activity in the visual word form area either because information failed to reach this area or because of a lack of attention to that brain region. Subsequently, other studies have confirmed the reduced conflict produced by posthypnotic suggestions in Stroop and flanker tasks (Augustinova & Ferrand, 2012; Parris, Dienes, & Hodgson, 2013). However, it should be noted that multiple routes could lead from a word to its meaning, thus it may be possible for a semantic effect to continue even if the word-form route was completely abolished. Therefore it remains somewhat uncertain whether posthypnotic suggestion works by abolishing input or by other means. In contrast, in one of our Stroop studies following IBMT, posterior brain potentials did not differ between the IBMT and relaxation control groups, but the frontal N2 (ACC as source), which has often been related to the effort to resolve conflict, was greatly reduced following IBMT. In addition, the subsequent late positive wave (P3) recorded over frontal electrodes was reduced in latency by IBMT. These findings suggested that conflict was resolved with less effort and more quickly and efficiently following meditation (Fan et al., 2015).

Although there are behavioral similarities between Stroop performance following IBMT or hypnotic suggestion, the underlying mechanisms may be different. Resting-state fMRI studies have shown reduced activation in the ACC during the hypnotic state (Deeley et al., 2012), while a highly active ventral ACC was found for IBMT participants in resting state (Tang et al., 2009). A possible interpretation of these findings is that the hypnotic state tends to reduce the mechanisms by which internal goals affect behavior though the executive attention network, leaving the ability of external input or working memory to control behavior intact. Thus performance in the Stroop task may be enhanced while the loss of effective internal goals allows the hypnotist to have a stronger control of behavior. Instead, meditation training seems to work by increasing self-control though strengthening activation and connectivity of the executive network. Therefore meditation enhances the ability of the control of behavior either through external or internal sources (Tang, 2017). Taken together, we propose that hypnosis and meditation change brain states and affect conflict resolution. Meditation increases resting-state activity within the ACC node of the executive attention or self-control network, whereas it appears that increasing depth of hypnosis reduces activity within the ACC. Hypnotic states may work to prevent internal goals from activating the executive attention or self-control network, leading to control by external input and prior instruction. In contrast, meditation training works to enhance executive attention and self-control to efficiently exert control and resolve conflict.

In summary, meditation is different from the psychedelic state and hypnosis in that meditation maintains self-control and self-awareness in an effortful or effortless way. In contrast, the psychedelic state and hypnosis are controlled by external psychedelic drugs and hypnotists, respectively. Who is in control seems to make the significant differences among these mental states and self-experiences supported by different brain mechanisms. Moreover, there are important phenomenological differences in the experiences of self-consciousness such as self-loss, self-dissolving, or ego dissolution induced by meditation, psychedelic state, and hypnosis. Current results indicate that self-consciousness is a multidimensional construct and can take several forms, which are more complicated than what we thought before. Future research is necessary to confirm and validate these hypotheses.

Potential side effects or adverse effects

Although meditation produces many positive effects on psychological health and cognitive function, some meditation participants sometimes may have negative experiences during meditation, such as fear, fatigue, or traumatic memories (Lindahl, Fisher, Cooper, Rosen, & Britton, 2017; Van Dam et al., 2018). For example, a recent study investigated meditation-related adversities in 60 Buddhist meditation practitioners and showed that negative side effects occurred for 12% of participants within the first 10 days of practice. Additionally, 25% encountered adversities while practicing for less than 1 h/day, and 30% had adverse experiences in daily practice (Lindahl et al., 2017). Although we do not know exactly what and how these individuals practiced meditation during the study period or what kind of techniques they used during the practices, these results suggest that simple practice can lead to adverse experiences. Even if such adverse effects occur only in 5% of participants, which is similar to the rate found in psychotherapy (Crawford et al., 2016), almost one million US adults per year may suffer from meditation-related negative side effects given that about 18 million US adults practice meditation annually (Clark, Black, Stussman, Barnes, & Nahin, 2015). In their review of methodological and conceptual issues in meditation research, Van Dam et al., suggested more attention should be paid to the adverse effects of meditation practices. Researchers should include questionnaires about adverse experiences during meditation practice, and better understand the nature of these experiences to determine whether these experiences are due to improperly following the instructions or due to the practice itself. It is highly likely that individuals with certain personality issues and/or psychiatric disorders (e.g., schizophrenia) should not engage in meditation practices, as their sensations, perceptions of self, and other metal processes are already disoriented in some ways, which could potentially be exacerbated by meditation practices.

However, there are also several common misunderstandings and incorrect practice strategies that may lead to some of these side effects. Based on our teaching experience, we would like to share a few simple and direct strategies that may help prevent and minimize these problems and adversities. Overall, our experience and observation suggest that participants who have "expectation and fear" about the practice and outcomes often have issues or side effects in meditation learning and practice. In contrast, having "no expectation and no fear" are related to positive experiences and outcomes.

Expectation refers to wanting certain goals or results to happen such as a specific practice feeling or experience, the strong desire to achieve a meditative state, or striving to obtain specific outcomes or benefits. These expectations would often interfere with the practice experiences since the individuals would likely have trouble focusing on their practice due to their constant awareness of their strong expectations. In reality, having goals in meditation is not a problem. However, the issue of adverse effects and experiences arises when unacceptance of any other feelings and outcomes occurs during the practice. For most of us, if things do not proceed as we expected, disappointment and frustration would naturally occur. Yet just like any other experiences in life, appropriately handling our expectations is critical for attaining our goals. For meditation practitioners, one of the helpful ways to think about meditation is that the process of meditation is the goal. There is no need to achieve certain goals during meditation practices because the present experiences are the goals. The key is that we are learning to maintain our focus and train our attention to be on the present-moment experiences, which is the goal of meditation. Additionally, the key to meditation is to accept any experiences with a nonjudgmental and nonreactive attitude, thus some experiences such as boredom and frustration (except for serious adverse experiences or sensations) could be opportunities for practitioners to observe. Having a proper understanding of meditation practices would reduce the likelihood of experiencing some adverse effects of meditation. According to Buddhism traditions, wanting always creates suffering because it is the cause of suffering. Without strong expectation for certain outcomes, one can naturally and easily settle down with the awareness itself, rather than fixate on certain unrealistic goals that interfere with the practice. Furthermore, wanting to control the mind often leads to failure and distress because we would never win this struggle as we are fighting with ourselves. Therefore some of the adversities during meditation are very likely due to this high expectation (Tang, 2017).

Another adverse experience is fear, which refers to avoidance (unwanting) of certain practice feelings and experiences. This is also a learned habit given that in daily life we often use avoidance or suppression strategies to hide our emotions and intentions in order to not to experience certain consequences. However, the intention to avoid a thought (suppression of unwanted thoughts) often leads to thinking about the very thought one hopes to suppress. In other words, what we resist persists. Fear is actually another form of expectation, that is, expectation and fear are two sides of the same coin

(Tang, Holzel, & Posner, 2015; Tang, 2017). Except for fear that relates to self-preservation or serious mental and physical discomfort, other forms of fear that happen during the meditation practice should be communicated to the instructors and potentially addressed using constructive strategies. Additionally, just like other adverse experiences, fear should also be better assessed through questionnaires and research should categorize different kinds of fear. For example, is it a fear of not obtaining benefit or a fear of physical and mental well-being? Relatedly, self-acceptance may also be associated with meditation-related adversities in that having a high self-acceptance would likely protect individuals from the pitfalls of perfectionism. Perfectionism and self-criticism seem to be good motivators in everyday life, but research has shown that people perform more effectively when motivated by encouragement, reward, and self-compassion. Exercises and practices of self-acceptance and self-compassion are important in meditation, which may be able to address some of issues associated with adverse effects and support us in finding optimal ways to motivate ourselves and achieve our goals naturally. Based on the idea of cause—consequence effects, we should be working on the cause rather than the outcomes, that is, working on the practice rather than expecting the outcomes, and gradually the outcomes may reveal themselves as the practice of meditation progresses.

In addition to the no expectation and no fear principle, another fundamental principle is "no joy no meditation." We all know that paying attention during meditation is a very challenging task and our common strategy is to devote more effort and try harder and harder to achieve focused attention. We have learned to put effort into achieving goals in the school or workplace, and more effort often means more success. Naturally, we would bring this mindset and approach to meditation learning and practice. For example, what we learned from a physical workout is to devote continuous effort when we aim to build muscles. Unfortunately, many people would feel upset when they seem to fail in achieving the mind-made goals after trying hard with more effort. Meditation works directly with our mind and brain, and we must follow the knowledge we learn from psychology and neuroscience in guiding our practices. First, self-control of our mind is important, but reward also drives our learning and behavior. Given that paying attention is not a natural reward, the control of mind or thoughts is not a reward-based learning and thus could not help our meditation practice and outcomes. Our translational research strongly suggests that joy is another driving factor of meditation. Without joy, our meditation becomes dry and boring, subsequently affecting our practice and behavior. Overall, based on our teaching experience of meditation and research on meditation practices, these principles and factors such as nonstriving and effortless control of attention and action, a sense of equanimity, curiosity, openness, and patience are critical for achieving a meditative state and reducing the likelihood of experiencing adverse effects related to meditation.

Last, but not least, the study of the individual differences in meditation is still in its infancy. We are grateful to have this opportunity to write this book on this important topic in the meditation field. We have tried our best to summarize the results and findings in the field, however, further questions still exist. For example, how to teach and disseminate meditation effectively? How to develop personalized meditation by considering genes, environment and experience for any individual? How to select and tailor certain experience in our life (e.g., meditation) to voluntarily shape our biological heredity and thus change our behavior, health, and quality of life? How to integrate meditation into daily activities of lifestyle? How to achieve a stable mind efficiently to cultivate wisdom, equanimity, and compassion? These important questions should be addressed with the help of new technology and scientific discovery in the future. However, we should keep in mind that given the limitations of the human mind, perspectives, and capacities as well as technological equipment, scientific research is always being constantly updated and, sometimes overturned. Therefore we should always be open-minded and mindful.

In this book, we share what we have learned from the research findings in the field and provide practical guidance for individuals who are interested in learning and practicing meditation. Our book is meant not as a complete summary of meditation research and application, but instead seeks to motivate further scientific and translational discussion. We sincerely hope this book contributes to the development of the field of meditation research and applications. We envision that with time, more new findings will advance our current understandings and knowledge of individual differences in meditation.

References

Anderson, T., & Farb, N. (2018). Personalizing practice using preferences for meditation anchor modality. *Frontiers in Psychology*, *9*, 2521.

Augustinova, M., & Ferrand, L. (2012). Suggestion does not de-automatize word reading: Evidence from the semantically based Stroop task. *Psychonomic Bulletin & Review*, *19*(3), 521−527.

Brandmeyer, T., & Delorme, A. (2013). Meditation and neurofeedback. *Frontiers in Psychology*, *4*, 688.

Bruijniks, S. J. E., DeRubeis, R. J., Hollon, S. D., & Huibers, M. J. H. (2019). The potential role of learning capacity in cognitive behavior therapy for depression: A systematic review of the evidence and future directions for improving therapeutic learning. *Clinical Psychological Science*, *7*(4), 668−692.

Burke, A. (2012). Comparing individual preferences for four meditation techniques: Zen, Vipassana (mindfulness), Qigong, and Mantra. *Explore: The Journal of Science and Healing*, *8*(4), 237−242.

Butler, S. F., & Strupp, H. H. (1986). Specific and nonspecific factors in psychotherapy: A problematic paradigm for psychotherapy research. *Psychotherapy: Theory, Research, Practice, Training*, *23*(1), 30.

Clark, T. C., Black, L. I., Stussman, B. J., Barnes, P. M., & Nahin, R. L. (2015). *Trends in the use of complementary health approaches among adults: United States, 2002–2012*. Hyattsville, MD: National Center for Health Statistics. (National Health Statistics Reports, 79).

Crawford, M. J., Thana, L., Farquharson, L., Palmer, L., Hancock, E., Bassett, P., . . . Parry, G. D. (2016). Patient experience of negative effects of psychological treatment: Results of a national survey. *British Journal of Psychiatry, 208*, 260–265.

Davidson, R. J., & Dahl, C. (2018). Outstanding challenges in scientific research on mindfulness and meditation. *Perspectives on Psychological Science: A Journal of the Association for Psychological Science, 13*, 62–65.

Deeley, Q., Oakley, D. A., Toone, B., Giampietro, V., Brammer, M. J., & Williams, S. C. R. (2012). Modulating the default mode network using hypnosis. *International Journal of Clinical and Experimental Hypnosis, 60*(2), 206–228.

Enriquez-Geppert, S., Huster, R. J., Scharfenort, R., Mokom, Z. N., Zimmermann, J., & Herrmann, C. S. (2014). Modulation of frontal-midline theta by neurofeedback. *Biological Psychology, 95*, 59–69.

Fan, Y., Tang, Y. Y., Tang, R., & Posner, M. I. (2014). Short term integrative meditation improves resting alpha activity and stroop performance. *Applied Psychophysiology and Biofeedback, 39*(3–4), 213–217.

Fan, Y., Tang, Y. Y., Tang, R., & Posner, M. I. (2015). Time course of conflict processing modulated by brief meditation training. *Frontiers in Psychology, 6*, 911.

Fox, K. C., Dixon, M. L., Nijeboer, S., Girn, M., Floman, J. L., Lifshitz, M., . . . Christoff, K. (2016). Functional neuroanatomy of meditation: A review and meta-analysis of 78 functional neuroimaging investigations. *Neuroscience & Biobehavioral Reviews, 65*, 208–228.

Fox, L. D., Rejeski, W. J., & Gauvin, L. (2000). Effects of leadership style and group dynamics on enjoyment of physical activity. *American Journal of Health Promotion, 14*(5), 277–283.

Frank, J. D. (1959). The dynamics of the psychotherapeutic relationship: Determinants and effects of the therapist's influence. *Psychiatry, 22*(1), 17–39.

Gruzelier, J. H. (2014). EEG-neurofeedback for optimising performance. I: A review of cognitive and affective outcome in healthy participants. *Neuroscience & Biobehavioral Reviews, 44*, 124–141.

Lemmens, L. H. J. M., van Bronswijk, S. C., Peeters, F., Arntz, A., Hollon, S. D., & Huibers, M. J. H. (2019). Long-term outcomes of acute treatment with cognitive therapy vs. interpersonal psychotherapy for adult depression: Follow-up of a randomized controlled trial. *Psychological Medicine, 49*, 465–473.

Lindahl, J. R., Fisher, N. E., Cooper, D. J., Rosen, R. K., & Britton, W. B. (2017). The varieties of contemplative experience: A mixed-methods study of meditation-related challenges in Western Buddhism. *PLoS One, 12*(5), e0176239.

Lippelt, D. P., Hommel, B., & Colzato, L. S. (2014). Focused attention, open monitoring and loving kindness meditation: Effects on attention, conflict monitoring, and creativity—A review. *Frontiers in Psychology, 5*, 1083.

Marzbani, H., Marateb, H. R., & Mansourian, M. (2016). Neurofeedback: A comprehensive review on system design, methodology and clinical applications. *Basic and Clinical Neuroscience, 7*(2), 143–158.

Millière, R. (2017). Looking for the self: Phenomenology, neurophysiology and philosophical significance of drug-induced ego dissolution. *Frontiers in Human Neuroscience, 11*, 245.

Millière, R., Carhart-Harris, R. L., Roseman, L., Trautwein, F. M., & Berkovich-Ohana, A. (2018). Psychedelics, meditation, and self-consciousness. *Frontiers in Psychology, 9*, 1475.

Oakley, D. A., & Halligan, P. W. (2013). Hypnotic suggestion: Opportunities for cognitive neuroscience. *Nature Reviews Neuroscience, 14*(8), 565–576.

Parris, B. A., Dienes, Z., & Hodgson, T. L. (2013). Application of the ex-Gaussian function to the effect of the word blindness suggestion on Stroop task performance suggests no word blindness. *Frontiers in Psychology, 4*, 647.

Penedo, F. J., & Dahn, J. R. (2005). Exercise and well-being: A review of mental and physical health benefits associated with physical activity. *Current Opinion in Psychiatry, 18*(2), 189–193.

Peterson, R. S., Smith, D. B., Martorana, P. V., & Owens, P. D. (2003). The impact of chief executive officer personality on top management team dynamics: One mechanism by which leadership affects organizational performance. *Journal of Applied Psychology, 88*(5), 795.

Preller, K. H., & Vollenweider, F. X. (2018). Phenomenology, structure, and dynamic of psychedelic states. *Current Topics in Behavioral Neurosciences, 36*, 221–256.

Raz, A., Fan, J., & Posner, M. I. (2005). Hypnotic suggestion reduces conflict in the human brain. *Proceedings of the National Academy of Sciences of the United States of America, 102*, 9978–9983.

Smigielski, L., Scheidegger, M., Kometer, M., & Vollenweider, F. X. (2019). Psilocybin-assisted mindfulness training modulates self-consciousness and brain default mode network connectivity with lasting effects. *NeuroImage, 196*, 207–215.

Tang, Y. Y. (2017). The neuroscience of mindfulness meditation: How the body and mind work together to change our behavior? London: Springer Nature.

Tang, Y. Y., Holzel, B. K., & Posner, M. I. (2015). The neuroscience of mindfulness meditation. *Nature Reviews Neuroscience, 16*, 213–225.

Tang, Y. Y., & Tang, R. (2014). Ventral-subgenual anterior cingulate cortex and self-transcendence. *Frontiers in Psychology, 4*, 1000.

Tang, Y. Y., Ma, Y., Wang, J., Fan, Y., Feng, S., Lu, Q., & Posner, M. I. (2007). Short term meditation training improves attention and self regulation. *Proceedings of the National Academy of Sciences, USA, 104*(43), 17152–17156.

Tang, Y. Y., Ma, Y., Fan, Y., Feng, H., Wang, J., Feng, S., & Fan, M. (2009). Central and autonomic nervous system interaction is altered by short term meditation. *Proceedings of the National Academy of Sciences, USA, 106*(22), 8865–8870.

Tang, Y. Y., & Posner, M. I. (2015). *Influencing conflict in the human brain by changing brain states. Hypnosis and meditation: Towards an integrative science of conscious planes.* Oxford University Press.

Truax, C. B., & Carkhuff, R. (2007). *Toward effective counseling and psychotherapy: Training and practice.* Transaction Publishers.

Van Dam, N. T., van Vugt, M. K., Vago, D. R., Schmalzl, L., Saron, C. D., Olendzki, A., ... Meyer, D. E. (2018). Mind the hype: A critical evaluation and prescriptive agenda for research on mindfulness and meditation. *Perspectives on Psychological Science: A Journal of the Association for Psychological Science, 13*(1), 36–61.

Williams, R., Farquharson, L., Palmer, L., Bassett, P., Clarke, J., Clark, D. M., & Crawford, M. J. (2016). Patient preference in psychological treatment and associations with self-reported outcome: National cross-sectional survey in England and Wales. *BMC Psychiatry, 16*(1), 4.

Index

Note: Page numbers followed by "*f*" refer to figures.

Printed in the United States
By Bookmasters